Seabirds in the North-East Atlantic

Climate Change Vulnerability and Potential
Conservation Actions

First edition

*Henry Häkkinen, Silviu Petrovan, Nigel G. Taylor,
William J. Sutherland and Nathalie Pettorelli*

OpenBook
Publishers

https://www.openbookpublishers.com

© 2023 Häkkinen et al.

ISBN Paperback: 978-1-80511-011-8
ISBN Hardback: 978-1-80511-012-5
ISBN Digital (PDF): 978-1-80511-013-2
DOI: 10.11647/OBP.0343

Cover image by Seppo Häkkinen (2023)
Cover design by Jeevanjot Kaur Nagpal

© Seppo Häkkinen

Contents

© Seppo Häkkinen

Introduction

0.1 What is this book?

This resource is part of a series produced by the Zoological Society of London and the University of Cambridge, which aims to (1) assess seabirds' vulnerability to climate change in the North-East Atlantic, and (2) identify potential conservation actions that could reduce this vulnerability. This guidance collates information from the scientific literature, non-governmental organisations' reports, conservation practitioner input and online databases into a single resource, and provides a reference manual to assist conservation planning. It is intended to be used by anyone who wishes to identify climate change threats to seabirds; to compare threats between different areas of the North-East Atlantic; to start a quantitative climate change vulnerability assessment for a local population; or to review options for conservation action in response to climate change.

This book synthesises available information for seabirds in the North-East Atlantic. The North-East Atlantic covers the OSPAR region of Europe, from the Barents Sea and Svalbard in the North, to the coast of Portugal in the South. We also included species and populations breeding in and around the Baltic Sea; this adjustment was made in response to known distributions of significant fish stocks, as well as information on areas known to be important breeding and/or wintering grounds for species otherwise common in Western Europe. We did not assess species or populations in the Mediterranean, around Greenland, around the Azores or around the Canaries.

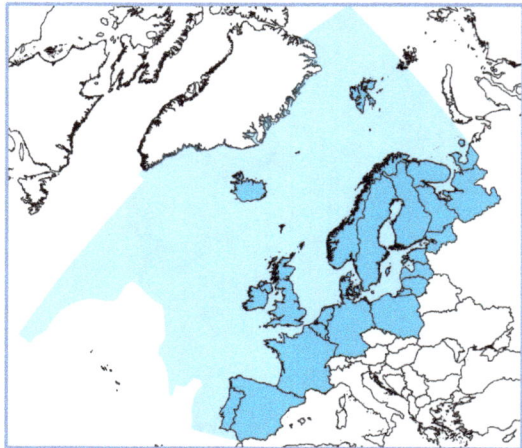

The area covered by this report (shaded in blue) based on the OSPAR region of the North Atlantic.

As part of planning and developing this series of resources, we consulted and collaborated with a variety of conservation stakeholders and practitioners from across Europe, and have added their knowledge and experience to complement the available published information. This includes currently unpublished impacts and first-hand experience on the practicality and effectiveness of conservation actions. We would like to thank everyone who contributed to the development of this work. This document is part of version 1.2, published in June 2023, but assessments may be updated based on feedback and newly available information. To check for updates to our assessments, please visit our website at: www.ZSL.org/seabird-guidelines and at https://doi.org/10.11647/OBP.0343.

0.2 What this book contains

This book contains two major sections. The first assesses how vulnerable seabirds in the North-East Atlantic are to climate change, and the second assesses the conservation actions available for each identified climate change related threat. We carried out an assessment for all seabird species that have a permanent breeding population in the North-East Atlantic.

A "seabird" is not a distinct taxonomic group, but is defined by any species of bird that predominantly relies on marine habitat for at least part of its annual cycle. We identified 48 species, loosely grouped as auks, cormorants, gannets, grebes, gulls, loons/divers, sea ducks, skuas and terns. There are several additional species that are marginally marine, or have at least a few populations that are marine, but in most cases we excluded such species as they are predominantly associated with terrestrial, freshwater or estuarine habitats. We use the English common names used as standard by Birdlife, though other synonyms may be more familiar to some readers. In particular, we use "loons" rather than "divers" and refer to "murres" rather than "guillemots". For further details please see the Birdlife Taxonomic Checklist (http://datazone.birdlife.org/species/taxonomy).

The following is a summary of each section of the guidance. For further information on how we compiled each section, please see our corresponding appendices that contain full references and information on sources used. For a full methodology, see the accompanying 'Methodology' folder in Appendix 2.

Section 1: Vulnerability to climate change. Section one reviews the vulnerability of each auk species to climate change, using the framework laid out by Foden et al., 2017. It subdivides vulnerability into three main categories:

- **Section 1.1: Exposure.** Exposure is a description of the nature, magnitude and rate of changes induced by climate change. We assessed exposure in four ways:

  ◦ **Section 1.1.1: Current impacts on seabirds attributed to climate change.**

This is a numbered list of the impacts of climate change on each species that has so far been observed in the North-East Atlantic. An impact is defined as a change in breeding success, abundance, survival, condition, behaviour or genetics that can be at least partially attributed to climate change. This includes: a) long-term trends in populations where climate change is believed to be a contributing driver, b) impacts of extreme climate events where the frequency/severity/duration of such events is known to be linked to climate change, c) an observed significant increase in exposure to a known threat (e.g. predators, parasites) where climate change is believed to be a contributing factor. Impacts may be: positive, where climate change has resulted in a positive change to a demographic parameter (e.g. breeding success or abundance), negative, where climate change has resulted in a negative change to a demographic parameter (e.g. breeding success or abundance), or neutral, where climate change has clearly had an effect on a population but it is unclear whether the effect is positive or negative (e.g. change in phenology with no recorded change in breeding success or abundance etc.). The location of these impacts is marked on the accompanying map by numbered icons. For a full list of sources see Appendix 1.1.1.

◦ **Section 1.1.2 Potential changes in breeding habitat suitability.** We here aim to predict how much of the species' current breeding range will be significantly less suitable in 2070-2100, based upon changes to the marine and terrestrial environment. We also estimate what proportion is likely to remain suitable, and whether parts of species' current ranges will become more suitable. The underlying species distribution model (SDM) considers predicted changes in temperature, precipitation, salinity, distance from the sea and marine chlorophyll concentration, as well as several species-specific variables which are detailed in the appendices. After comparing estimated habitat suitability between 2020 and 2070-2100, we split the coastal region of North-West Europe into one of the four following categories: 1) Habitat is currently suitable for a given species but will likely become significantly less so in the future (marked in red on the map), 2) Habitat is currently suitable and will likely remain stable in the future (marked orange on the map), 3) Habitat is currently suitable but will become significantly more so in the future (marked green on the map) and 4) Habitat is not currently suitable and will not be in the future (not marked on the map). There is considerable uncertainty around these estimates, and as such they should be understood as an indicator of risk rather than a firm prediction. Note that maps are aggregated and enlarged to make small islands more visible and are not exact representations of species ranges.

◦ **Section 1.1.3: Predicted changes in key prey species.** For each species we compiled a list of key marine prey species, as well as existing estimates of how their range and abundance may change between now and 2100. We identified areas where one or more key prey species are likely to become significantly less common in the future and highlight these as areas of high risk.

  ◦ **Section 1.1.4: Climate change impacts outside of Europe.** In some cases climate change is known to impact populations outside of our study area. These data provide supporting evidence for impacts in Europe, highlighting impacts that may be of concern to populations in the future, even if those impacts have not so far been observed in the North-East Atlantic. In selected cases, we summarise the nature of the impacts and the general area in which they occur. Further details and references are provided in Appendix 1.

• **Section 1.2: Sensitivity.** Sensitivity is the degree to which a species is likely to be affected, either adversely or beneficially by climate change. Sensitivity is expected to be shaped by species traits (e.g. body size, home range area or sociality) and is determined largely by intrinsic, biological features that have evolved over time. We used a list of candidate traits based on Foden & Young (2016) and identified which, if any, each species possesses.

• **Section 1.3: Adaptive capacity.** Adaptive capacity is the potential, capability, or ability of a species to adjust to climate change, to moderate potential damage, or to respond to the consequences. This may be either through changes in behaviour or changes in physiology. We used a list of candidate traits based on Foden & Young (2016) and identified which, if any, each species possesses.

Section 2: Potential conservation actions.

In this section we list potential conservation actions in response to climate change impacts and the evidence behind their effectiveness. For each impact we have compiled a list of local actions that may prevent or limit the direct or indirect impacts of climate change. Potential conservation actions were compiled from Conservation Evidence (Williams et al. 2013) as well as supplementary literature searches of published seabird conservation studies up to July 2021, and from direct consultation with practitioners.

By "local action" we mean conservation actions that directly prevent or limit an impact, and act on a local population scale. While broader scale action tackling climate change and ecosystem scale conservation are incredibly valuable, we intend this resource to be used as a guide to help conservation of populations at a local level. See the "making evidence-based decisions and how to use this guidance" section for further information.

We do not include actions that aim to increase the resilience of seabird populations to climate change by reducing other impacts (e.g. legal protection of species, hunting bans, reducing pollution). In some cases where very few viable direct actions are available, or likely to be effective, we include some discussion of indirect actions to support populations. However, indirect actions are often part of complex cause-and-effect pathways, and it is very difficult to assess their overall effectiveness on conserving seabird populations.

By "direct impact" we mean the direct physical impacts of climate change, or related changes in the physical environment, on seabirds. Examples would be heat stress caused by rising temperatures or increased physiological costs of foraging due to stronger winds. By "indirect impact" we mean changes in ecological processes that then impact seabirds. Examples would be changes in prey range, abundance or composition, increase in predation due to range-shifts, or changes in disease prevalence.

We do not include actions in response to human activities, even if the distribution or intensity of these might be influenced by climate change. For example, renewable energy infrastructure is likely to change in response to climate change, and is likely to increase exponentially in future decades to tackle the climate crisis. However, as this is a human-mediated impact, it is not included in this guidance.

For this section we group by climate change impact rather than species. For example, if multiple species are likely to suffer prey shortages in the breeding season, we summarise the possible actions in response for all species in the group at once. If actions and evidence are specific to one or a few species, this is discussed in the action summary and footnotes.

For each action we assessed the available evidence about its effectiveness and the relative strength, relevance and transparency of the supporting evidence (on a scale from 1 to 5). The following points give an outline of the criteria used to assess each score; for a full methodology see Appendix 2. For each action, we also provide a list of references and sources we used in Appendix 2.

- **Effectiveness:** For each action we assessed how effective it was when carried out on seabirds. Each conservation action is rated based on the evidence for its effectiveness, ranging from "likely to be beneficial" to "likely to be harmful". Effectiveness categories are taken from Conservation Evidence (Williams et al. 2021), and thus pertain to all birds unless noted otherwise. For a full methodology see Appendix 2. Studies documenting actions' effectiveness, specifically on seabirds, were used as the primary evidence base, but if a suggested action has not been trialled on seabirds we also consulted available evidence based on conservation studies targeting other bird species. We also included information from practitioners (if available) regarding an action's effectiveness or practicality for key populations in Europe.

- **Strength:** This refers to several characteristics of the underlying evidence base regarding the relative robustness and coverage of the evidence. In particular, it is based on how many studies have explored a given conservation action, did they test it on a large number of individuals or have a large number of replicates, has it been tested in various parts of a species' range, and did the authors have a clear and sensible metric for success and was it measured robustly.

- **Relevance:** This refers to how much of the underlying evidence base is composed of evidence specifically regarding the species group in question (e.g. auks). If an action is rated as beneficial, then the relevance score refers to how confident practitioners can be that a given conservation action is beneficial specifically for the focal species group.

- **Transparency**: This refers to how much of the underlying evidence base is composed of evidence that has clear methodology, readily available and detailed data, and a clear, evidence-based rationale, all of which has preferably been peer-reviewed.

Appendices

As an evidence-based guidance resource, being clear about where the information underlying our assessment has come from is key. Therefore, for each of the sections in the main text there is a corresponding appendix section containing references, additional detail or notes on methodology for those who wish to examine the primary sources or find additional reading. Appendix 1 contains additional information for Section 1 of the guidance, and Appendix 2 contains additional information for Section 2. Subheadings in the appendix match those in the main text. For example, if you would like to find the sources we used to create Section 1.1.1 (Current impacts attributed to climate change), then please consult Appendix 1.1.1.

0.3 What this book does not contain

A relative assessment of risk or effectiveness. Different populations face different combinations of risks and to different degrees of severity. For this assessment it was not possible to assess or rank the greatest threats to each population. Instead, we list all identified factors that contribute to vulnerability and the evidence behind them. Practitioners can however use this guide as a starting point to assess threats posed by climate change to their local population.

Recommendations for specific courses of action. This guidance is intended for use by practitioners as a reference guide to highlight threats and potential conservation actions for a given species. What action is most appropriate in a given scenario is dependent on many different factors, including ecological, financial, political and social concerns. This guidance should be considered in addition to the experience and judgement of those who work in the field.

0.4 Making evidence-based decisions and how to use this guidance

This guidance is a resource for a much wider framework, namely assessing threats to biodiversity and carrying out evidence-based conservation action. There are several published decision-making frameworks for conservation, and we will provide an example here based on the evidence-to-decision tool (https://evidence2decisiontool.com/), which identifies three major steps to making decisions. Here, we provide a brief summary of these steps and then detail where and how this guidance is intended to facilitate this process.

1) **Define the decision context.** What is being targeted, a specific site, a species, a population, or other? Is there a threat that needs to be addressed? If there are multiple threats, which should be addressed first? Which threat is the most urgent, should it be addressed first as a priority? How much impact could this threat have? What are the relevant ecological, physical, socio-economic and cultural factors that may be beneficial or a barrier for conservation? What are the goals of the conservation effort (this may include short-term and long-term goals such as: successfully moving nests, decreasing number of nests destroyed, increasing fledgling success rate, increasing population size over a decade)?

2) **Gather evidence.** If action is needed in response to identified threats, what are the potential actions? What is the evidence behind these actions; which are known to be effective? Is this action likely to be effective when applied to the specific situation at hand? Is an action feasible to carry out at the scale required? What are the financial and physical resources needed? What are the risks, costs and benefits of each action? Is it possible there will be unintended consequences? Is the action acceptable to stakeholders (e.g. will the action negatively impact another conservation target)?

3) **Make an evidence-based decision** Which threats need to be addressed? Which actions should be carried out? What is the justification for these decisions? After carrying out stages 2 and 3, it might be that the primary threat is extremely difficult to address, so is there another way to support a population by addressing another, secondary threat? Should the action instead focus on supporting and protecting the population in another way? How will you document and report the

conservation project? How will you monitor and evaluate the success of the action?

This guidance book provides key information for steps 1 and 2 of the above framework. **1)** It aims to help practitioners identify current and future threats to seabird species from climate change, and where these are likely to be most pressing. This guidance focuses on species-level context and identifies ecological and physical factors (through sensitivity and adaptive capacity), that are major barriers and opportunities for species to adapt. **2)** This guidance lists potential actions in response to identified climate change impacts. Practitioners can review these potential actions, and assessments of their effectiveness. While it's not possible to provide site-specific context, we have also included some information on acceptability and feasibility based on practitioner experience and feedback.

When combined with practitioner experience and judgement, this guidance should assist decisions regarding how to (a) prioritise species and areas for conservation, and (b) make an evidence-based decision on if and how active intervention should be carried out.

0.5 References

Foden, W. B., et al. "Climate change vulnerability assessment of species." WIREs Clim Change 10.1 (2019): e551.

Williams, D. R., et al. "Bird conservation: global evidence for the effects of interventions". Vol. 2. Pelagic Publishing (2013).

Williams, D. R., et al. "Bird Conservation." Pages 137-281 in: W.J. Sutherland et al. (Eds.) What Works in Conservation 2021. Open Book Publishers, Cambridge, UK (2021).

0.6 License

0.7 Preferred citation

Häkkinen, H., Petrovan, S., Taylor, N. G., Sutherland, W. J., Pettorelli, N. "Seabirds in the North-East Atlantic: Climate Change Vulnerability and Potential Conservation Actions" Open Book Publishers (2023): 1-278. DOI: 10.11647/OBP.0343

0.8 Contributors

We wish to thank the following people for their input on this project, without their knowledge, experience and feedback this project would not have been possible: Jón Aldará; Hany Alonso; Orea Anderson; Tycho Anker-Nilssen; Helder Araújo; José Manuel Arcos; Christophe Aulert; Bryony Baker; Elmar Ballstaedt; Rob van Bemmelen; Daniel Bengtsson; Richard Berridge; Julie Black; Daisy Burnell; Bernard Cadiou; Kees Camphuysen; Letizia Campioni; Rodrigo Martínez Catalán; Beth Clark; Nina Dehnhard; Maria Dias; Leonie Enners; Mats Eriksson; Javi Franco; Morten Frederiksen; Bob Furness; Maria Gavrilo; Ros Green; Gunnar Thor Hallgrimsson; Sjúrður Hammer; Erpur Snær Hansen; Sveinn Are Hansen; Martti Hario; Stephen Hurling; Linnet Jessell; Mark Jessop; Rebecca Jones; Birgit Kleinschmidt; Urša Koce; Yann Kolbeinsson; Joris Laborie; Adrien Lambrechts; Meelis Leivits; Raphaël Leprince; Ulrik Lötberg; Klaudyna Maniszewska; Mike Meadows; Szabolcs Nagy; Steffen Oppel; Ana Payo-Payo; Kjeld Tommy Pedersen; Daniel Piec; Jaime Ramos; Frederic Robin; Aðalsteinn Örn Snæþórsson; Iben Hove Sørensen; Antra Stipniece; Sophie Thomas; Danni Thompson; Antonio Vulcano; Eric Walter; Ilka Win; Ramunas Žydelis.

In addition we would like to thank the EAZA Charadriiformes TAG for their input and contributions regarding seabird treatment, rehabilitation and ex-situ breeding programmes. We would particularly like to thank Simon Matthews for leading this collaboration.

0.9 Funding

We would like to thank our funders for their support during this project. This work is funded by Stichting Ave Fenix Europa. NP is funded by Research England. SP and WJS are funded by Arcadia, The David and Claudia Harding Foundation and MAVA.

© Seppo Häkkinen

Auks
(Alcidae)

An assessment of climate change vulnerability and potential conservation actions for auks in the North-East Atlantic

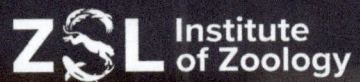

UNIVERSITY OF CAMBRIDGE

ZSL Institute of Zoology

https://doi.org/10.11647/OBP.0343.01

1 Razorbill *(Alca torda)*

1.1 Evidence for exposure

1.1.1 Potential changes in breeding habitat suitability (by 2100):

■ Current breeding area that is likely to become less suitable (80% of current range).

■ Current breeding area that is likely to remain suitable (18%).

■ Current breeding area that is likely to become more suitable (2%).

1.1.2 Current impacts climate change:

① **Negative Impact:** Extreme storms during the razorbill breeding season have led to wide-spread nest destruction, nesting failure and a net reduction in annual population production.

② **Negative Impact:** As sea temperatures have increased over time, razorbill productivity has decreased, most likely due to changes in prey availability.

③ **Neutral Impact:** Key prey species have shifted their life-cycle, likely in response to climate change, but razorbills have not adjusted in response. There is concern this could result in trophic mismatch, but no overall effect on breeding success has so far been observed.

1.1.3 Predicted changes in key prey species:

④ Key prey species are likely to decline in abundance in the Baltic, the Irish Sea and the English Channel.

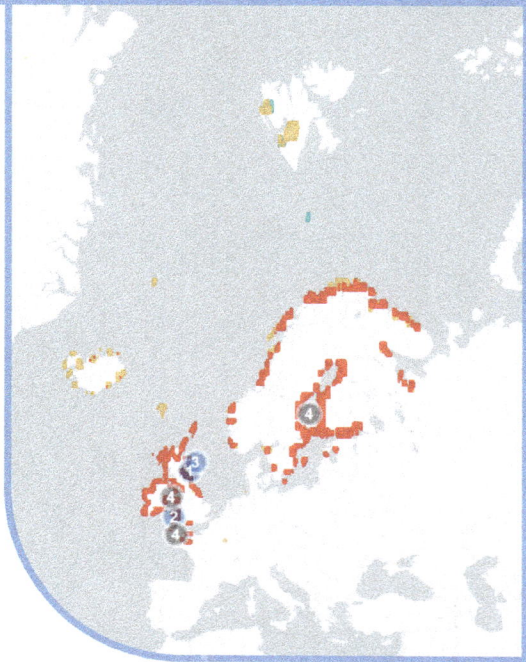

1.2 Sensitivity

• Razorbills prefer to nest in exposed places, which makes nests particularly vulnerable to storms. Any increase in the frequency or intensity of storms during the breeding season is likely to have severe consequences on razorbill breeding success.

• Razorbill survival in many areas correlates with sea surface temperature, likely due to changes in prey abundance. Projected temperature increases are therefore likely to decrease razorbill survival during the non-breeding season.

• Razorbills are prone to mass-mortality events ("wrecks"), both across Europe and in North America. The exact cause behind them is not certain, but likely related to food availability and winter storms. This makes drastic population reductions more likely, as well long recovery periods.

• Many razorbill populations are heavily reliant on a few prey species, especially during the breeding period. Any change in prey availability, particularly sandeels, is likely to have consequences for razorbill populations.

• Razorbills are vulnerable to mammal predation, and the spread of introduced mammals (favoured by climate change) could threaten more northern populations than previously.

• This species has a long generation length (>10 years), which may slow recovery from severe impacts and increases population extinction risk.

1.3 Adaptive capacity

• Razorbills are pelagic and are easily capable of reaching new potential colony sites. In North America there is some evidence they will colonise new areas to match key prey species ranges. However, there is no observed example of this occurring in Europe.

Razorbill © Seppo Häkkinen

2 Little Auk (*Alle alle*)

1.1 Evidence for exposure

1.1.1 Potential changes in breeding habitat suitability (by 2100):

■ Current breeding area that is likely to become less suitable (94% of current range).

■ Current breeding area that is likely to remain suitable (6%).

■ Current breeding area that is likely to become more suitable (0%).

1.1.2 Current impacts attributed to climate change:

1 Negative Impact: Warmer temperatures correlate with longer foraging trips and lower little auk productivity, most likely due to decreased prey availability.

2 Neutral Impact: Little auks are breeding earlier in correlation with warmer temperatures, so far no negative consequence has been observed.

3 Neutral Impact: Extreme storms during the non-breeding season have led to mass mortality of little auks ('wrecks').

1.1.3 Predicted changes in key prey species:

No key prey assessment was carried out for this species.

1.1.4 Climate change impacts outside of Europe:

• Loss of sea ice and availability of new prey items due to climate change has led to increased little auk breeding success in Greenland.

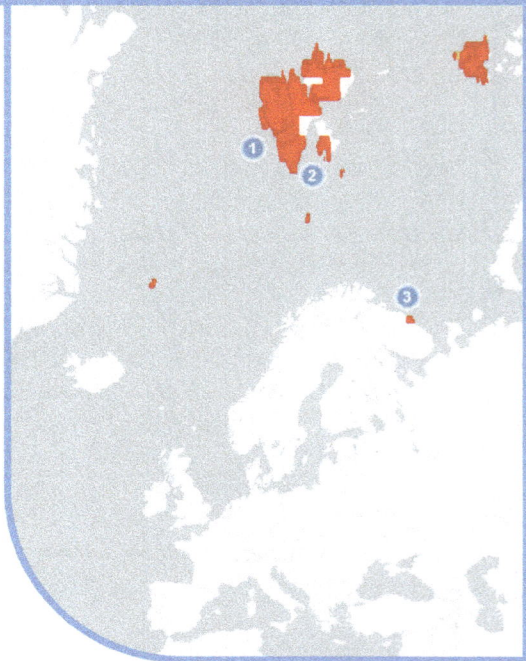

1.2 Sensitivity

• Little auks expend a great deal of energy in flight, and their chicks also have very high energy demands. This means that auks must focus on high-energy prey and are sensitive to even small changes in high-energy prey availability.

• Auk colonies that nest near warmer seas typically have longer and less successful foraging trips for nesting auks, which ultimately lowers chick growth and survival. Populations that therefore nest in areas with projected increases in sea temperature would likely be negatively impacted by climate change.

• Little auks on Svalbard may rely on environmental signals to time breeding events, in particular the timing of snow melt. This theory is not confirmed, but is supported by recent observed changes in phenology. If there is a strong reliance on such environmental cues, it may lead to trophic mismatch with prey species in the future.

• Little auks are dependent on a few species of copepods throughout the year. Any changes in availability or range of these species is likely to have a significant impact on little auks.

• Little auks congregate in large numbers in a relatively small area in the non-breeding season in the Atlantic. Any negative impact in this area is likely to have severe consequences for little auk populations.

• Little auks are known to be sensitive to climate change, and have suffered regional extinctions. They have previously suffered range contractions and local extinctions in the 19th-20th century most likely because of global warming (though historical changes were not primarily anthropogenic). As a result, it has lost much of its previous range in Iceland and Greenland.

• Little auk populations are large, but many are poorly monitored and exposed to multiple potential pressures. Impacts may be difficult to identify and conservation is likely to be difficult to carry out.

1.3 Adaptive capacity

• Little auks in Svalbard and Greenland have plastic foraging behaviour that has helped them compensate for local changes in prey availability and sea ice. However this plasticity likely has a limit and projections suggest they can only compensate to a certain extent.

• Little auks are thought to show strong fidelity to breeding sites. This may reduce the ability of little auks to adapt to changes in local conditions, as they are unlikely to change breeding sites in response to change.

3 Black Guillemot *(Cepphus grylle)*

1.1 Evidence for exposure

1.1.1 Potential changes in breeding habitat suitability (by 2100):

🟥 Current breeding area that is likely to become less suitable (76% of current range).

🟨 Current breeding area that is likely to remain suitable (24%).

🟩 Current breeding area that is likely to become more suitable (0%).

1.1.2 Current impacts attributed to climate change:

① **Negative Impact:** Heavy rainfall events and high water level has led to flooding of nests and lower hatching success in the Baltic. Such flooding events are becoming more common, and are likely to further increase. Debris left by storms and flooding can also make large areas of shoreline less suitable for breeding.

② **Negative Impact:** Range expansion of American mink, partly assisted by climate change, has led to increased rates of predation at guillemot colonies.

③ **Neutral Impact:** Guillemots have shifted their laying date, most likely linked to an increase in sea surface temperature and prey availability.

1.1.3 Predicted changes in key prey species:

④ Key prey species are likely to decline in abundance in the Irish Sea and the south-west coast of Sweden.

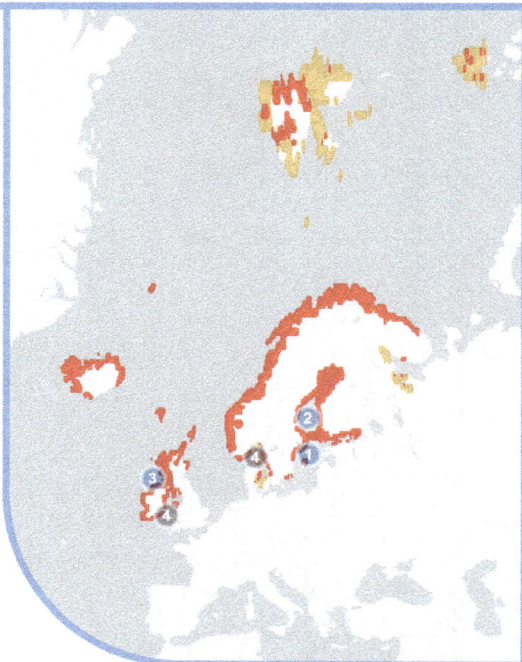

1.2 Sensitivity

• Black guillemots are particularly vulnerable to predation from mink and rats, due in part to their nesting on the ground in exposed areas. Range expansion of mammalian predators due to climate change (which is already occurring in Scandinavia) may have large impacts on guillemot populations.

• Black guillemots often nest in exposed areas close to water level, which make them vulnerable to flooding either from sea-level rise, extreme precipitation or tidal surges. If climate change results in a higher frequency of any of these events during the summer, it could severely impact black guillemot breeding populations.

• Black guillemots in Europe are declining, though with highly variable severity. There are various underlying causes including marine pollution, predation by invasive species, by-catch, disruption by wind farms and hunting. Any additional pressure from climate change is likely to accelerate these declines.

1.3 Adaptive capacity

• Black guillemots can shift their phenology in response to changes in environmental conditions. They may therefore be able to adjust to changes in conditions and prey availability and therefore mitigate impacts of climate change.

• In recent decades black guillemots have established several colonies in new areas around the Irish Sea and North America, which suggests they may be capable of range shifts in response to climate change.

© Seppo Häkkinen

4 Atlantic Puffin *(Fratercula arctica)*

1.1 Evidence for exposure

1.1.1 Potential changes in breeding habitat suitability (by 2100):

■ Current breeding area that is likely to become less suitable (68% of current range).

■ Current breeding area that is likely to remain suitable (31%).

■ Current breeding area that is likely to become more suitable (1%).

1.1.2 Current impacts attributed to climate change:

1 **Negative Impact:** Changes in puffins' prey availability during breeding season has led to decreased breeding success.

2 **Negative Impact:** Changes in puffins' prey availability during non-breeding season has led to increased mortality and population declines.

3 **Negative Impact:** Changes in vegetation has led to fewer suitable puffin nest-sites.

4 **Negative Impact:** Extreme storms during the non-breeding season have led to mass-mortality of puffins ('wrecks').

5 **Neutral Impact:** Puffins have changed their wintering range.

1.1.3 Predicted changes in key prey species:

6 Key prey species are likely to decline in abundance in the Irish Sea and the English Channel.

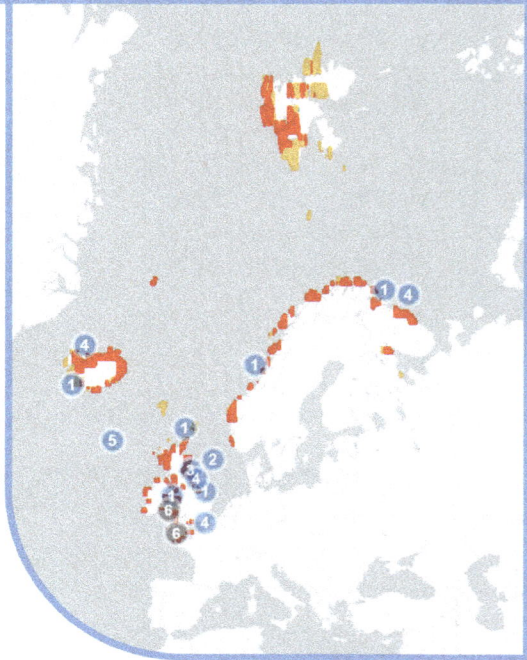

1.1.4 Climate change impacts outside of Europe

• Some colonies in North America have changed their laying phenology, presumably in response to temperature and/or prey availability.

1.2 Sensitivity

• Puffins are prone to population crashes, typically either from lack of prey during the breeding season or the effects of winter storms. This makes drastic populations reductions more likely, as well as long recovery periods.

• Adult puffin survival can drop sharply when there is an increase in the frequency, duration and intensity of winter storms, most likely due to increased foraging difficulty. Note: there is conflicting evidence regarding the effect of storms on puffins. However, puffins in areas most prone to extreme weather are more likely to be severely affected.

• Puffin colony success across Europe is correlated to copepod abundance, as they support many fish populations. In recent decades many areas in the south of the north-east Atlantic have become less suitable for copepods and this trend is likely to continue in the future. A decrease or range shift in copepods will likely have severe impacts on seabird colonies in the north-east Atlantic.

• This species has a long generation length (>10 years), which may slow recovery from severe impacts and increases population extinction risk.

• Puffins are declining rapidly in many parts of Europe, which has most major populations globally. Any additional pressure from climate change is likely to accelerate these declines.

1.3 Adaptive capacity

• Puffins travel long distances and could theoretically reach areas suitable for new colonies. However, while there have been examples of puffins colonising new areas (with or without human assistance), in general they have high site fidelity and rarely colonise new areas.

• Recent observations in Iceland report that puffins have started to swap prey species at some colonies, especially where major prey species have declined. The extent of this switch and the consequences are currently unknown.

• Colonies of puffins on Farne Islands and on Isle of May are breeding later, but not in correlation to changes in sea temperature. This may be due to other environmental changes in breeding or non-breeding areas. While this shift may be adaptive, it may also result in trophic mismatch if breeding cues don't match prey availability.

5 Common Murre *(Uria aalge)*

1.1 Evidence for exposure

1.1.1 Potential changes in breeding habitat suitability (by 2100):

■ Current breeding area that is likely to become less suitable (66% of current range).

■ Current breeding area that is likely to remain suitable (31%).

■ Current breeding area that is likely to become more suitable (3%).

1.1.2 Current impacts attributed to climate change:

① **Negative Impact:** High-wind events in the non-breeding season have led to mass mortality of murres in recent years.

② **Negative Impact:** Extreme storms during the non-breeding season have led to mass mortality of murres ('wrecks').

③ **Negative Impact:** More frequent extreme storms during murres' breeding season has increased foraging difficulty and reduced food fed to chicks.

④ **Negative Impact:** Extreme storms during murres' breeding season have led to wide-spread nest destruction, nesting failure and a net reduction in annual population production.

⑤ **Negative Impact:** Changes in murres' prey availability during the breeding season has led to decreased breeding success.

⑥ **Neutral Impact:** Murres are more likely to skip breeding in warmer weather, and this behaviour is becoming more frequent. While this is a cause for concern, it is unclear what effect this will have on the population in the long-term.

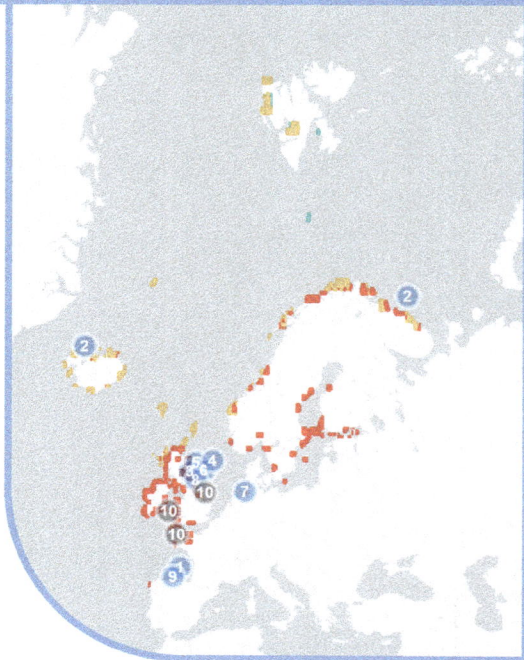

7 **Negative Impact:** Heatwaves have resulted in significant murre chick mortality. The frequency and severity of heatwaves is likely to increase.

8 **Neutral Impact:** Common murres have changed their phenology, potentially in response to climate change but the mechanism is unclear.

9 **Positive Impact:** A shift towards warmer, drier and calmer conditions has correlated with higher population abundance. Mechanism unknown, but likely mediated through prey availability and lower energetic costs.

1.1.3 Predicted changes in key prey species:

10 Key prey species are likely to decline in abundance in the Baltic, the Irish Sea and the English Channel.

1.2 Sensitivity

• Murres are prone to sporadic mass-mortality events ("wrecks"), both across Europe and in North America. The likely cause varies between wrecks, ranging from summer heatwaves, to prolonged extreme wind events in winter. Changes in extreme weather are likely to have significant effects on murre mortality.

• Murres in the Baltic are dependent on sprat as a food source and studies suggest they have limited ability to switch to other prey items. Declines in sprat due to climate change would likely have a severe impact on murres.

• Murre colony success across Europe seems to be tied to copepod abundance, as they support many fish species. In recent decades many areas in the south of the north-east Atlantic have become less suitable for copepods and this trend is likely to continue in the future. A decrease or range shift in copepods will likely have severe impacts on seabird colonies in the north-east Atlantic.

• For multiple populations, there is a strong correlation between breeding success and sea temperature in common murres; populations are negatively affected by warmer temperatures, most likely due to changes in marine ecosystems and prey availability. Projected warming is therefore likely to negatively impact many murre populations.

• This species has a long generation length (>10 years), which may slow recovery from severe impacts and increases population extinction risk.

• While most populations in Europe are stable, some are declining rapidly (notably in Iceland). Any additional pressure from climate change is likely to accelerate these declines.

1.3 Adaptive capacity

• There is evidence murres can change their phenology in response to temperature changes. Some populations have shifted their laying date in correlation with temperature changes, others have not. Other populations have changed their arrival date after migration, but not their laying date.

• There appears to be a weak genetic structure across colonies, which could be a result of high dispersal between colonies. This may increase resilience and aid population recovery following wrecks.

• In contrast to murres in the Baltic (see Sensitivity section), studies on populations in the North Sea and eastern Canada have found that murres can switch prey and spend more time at sea to compensate for changes in prey availability.

Common murres © Seppo Häkkinen

© Seppo Häkkinen

6 Thick-billed Murre *(Uria lomvia)*

1.1 Evidence for exposure

1.1.1 Potential changes in breeding habitat suitability (by 2100):

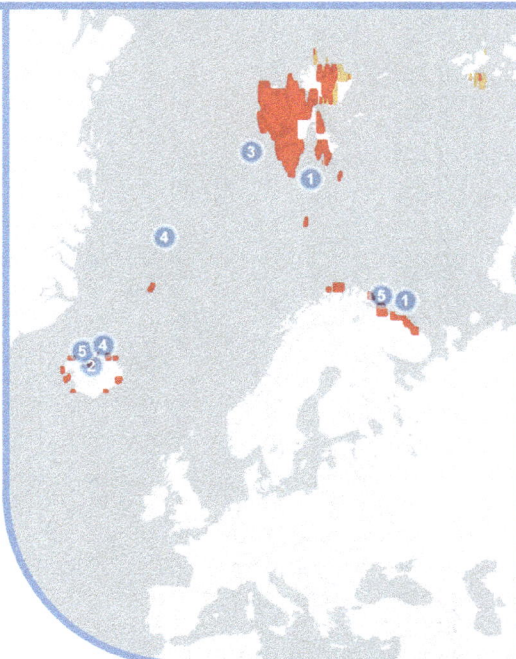

■ Current breeding area that is likely to become less suitable (87% of current range).

■ Current breeding area that is likely to remain suitable (13%).

■ Current breeding area that is likely to become more suitable (0%).

1.1.2 Current impacts attributed to climate change:

① **Negative Impact:** Changes in thick-billed murres' prey availability during the non-breeding season has led to increased mortality.

② **Negative Impact:** Changes in thick-billed murres' prey availability during the breeding season has led to decreased breeding success.

③ **Neutral Impact:** Changes in thick-billed murres' prey availability during the breeding season has led to increased mortality.

④ **Neutral Impact:** Thick-billed murre populations are typically smaller and decline in areas with increasing sea temperatures. Mechanism unclear.

⑤ **Neutral Impact:** Extreme storms during the non-breeding season have led to mass mortality of murres ('wrecks').

1.1.3 Predicted changes in key prey species:

No key prey species are predicted to decline for this species.

1.1.4 Climate change impacts outside of Europe:

• Thick-billed murres are known to be impacted by climate change outside of Europe. Impacts include increased predation by polar bears, increased parasitism by mosquitoes (leading to breeding failure), and increased mortality caused by algal blooms.

• Changes in the marine ecosystem in the Canadian high Arctic, driven by climate change, has resulted in higher concentrations of mercury bioaccumulated in thick-billed murres. No long-term impact on population health has been observed so far.

1.2 Sensitivity

• This species has a long generation length (>10 years), which may slow recovery from severe impacts and increases population extinction risk.

1.3 Adaptive capacity

• While a wide-roaming pelagic species, thick-billed murres have very high site fidelity. Moreover, juvenile dispersal is also very low. It is unlikely that murres will establish new colonies in response to climate change.

© Seppo Häkkinen

Potential actions in response to climate change: Auks (Alcidae)

In this section we list and assess possible local conservation actions that could be carried out in response to identified climate change impacts. This section is not grouped by species, but by identified impacts. If an impact or action is specific to one or a few species, this information is included in the action summary or in the footnotes. Effectiveness, relevance, strength and transparency scores are based on the available evidence we collated (see Appendix 2), and therefore all statements regarding limited or a lack of evidence relate to the collated evidence base, and does not infer that no such studies exist.

1 Impact: Increase in mammal predation

Summary:

Invasive mammals are a major threat to many seabird populations, and as such there is a well-established literature on mammal exclusion, management and eradication detailing effective methods and case studies. However, there are more limited options when the mammalian predator in question is itself a conservation target, or is not easily managed. Nevertheless, for many situations there are several, well-researched, actions available that can benefit seabird populations effectively.

Intervention	Evidence of Effectiveness	R	S	T
Manage/ eradicate mammalian predators	There are numerous examples of benefits to seabirds, including auks, following mammal eradication, though this depends on the effectiveness of the methods used and the species in question. Control of rodents and mustelids are particularly well studied.	3	5	3
Physically protect nests with barriers or enclosures	Has been successfully trialled in numerous ground-nesting seabird species. However, we found no studies that look at effectiveness for any auk species.	2	4	4

		R	S	T
Reduce predation by translocating predators	Few trials on seabirds, and only one on auks (*Synthliboramphus hypoleucus*). Existing evidence suggests this action can be beneficial and reduce egg/chick predation, and could be a possible action if other forms of predator management are not viable.	2	4	3
Repel predators with acoustic, chemical or visual deterrents	This is a hypothetical action. We found no published studies assessing this action's effectiveness for seabirds.	NA	NA	NA
Use supplementary feeding to reduce predation	Very few trials on seabirds, and none on auks. No studies have shown this action is effective.	1	4	3

Green = Likely to be beneficial. Red = Unlikely to be beneficial, may have negative impact. Orange = contradicting or uncertain evidence. Grey = Limited evidence.
R = relevance rating. S = strength rating. T = transparency rating. All ratings on a scale of 1 to 5, where 5 is the highest.

Details:

Manage/eradicate mammalian predators
Relevance (R): 6 studies in the evidence base focus on auks, 40 on other seabirds and 3 on other birds. **Strength (S):** The evidence base was comprised of 52 studies. Of these 44 were considered to have a good sample size, and 34 had a clear metric for effectiveness. **Transparency (T):** 52 studies included were published and peer-reviewed, of which 5 were literature reviews or meta-analyses, 0 were from the grey literature, and 0 were anecdotal. Of the studies included, 24 had a published methodology, and 28 justified their rationale.

Physically protect nests with barriers or enclosures
Relevance (R): 0 studies in the evidence base focus on auks, 12 on other seabirds and 6 on other birds. **Strength (S):** The evidence base was comprised of 18 studies. Of these 16 were considered to have a good sample size, and 12 had a

clear metric for effectiveness. **Transparency (T):** 17 studies included were published and peer-reviewed, 0 were from the grey literature, and 0 were anecdotal. Of the studies included, 11 had a published methodology, and 12 justified their rationale.

Reduce predation by translocating predators

Relevance (R): 1 study in the evidence base focusses on auks, 1 on other seabirds and 2 on other birds. **Strength (S):** The evidence base was comprised of 4 studies. Of these 4 were considered to have a good sample size, and 3 had a clear metric for effectiveness. **Transparency (T):** 4 studies included were published and peer-reviewed, 0 were from the grey literature, and 0 were anecdotal. Of the studies included, 2 had a published methodology, and 3 justified their rationale.

Use supplementary feeding to reduce predation

Relevance (R): 0 studies in the evidence base focus on auks, 1 on other seabirds and 3 on other birds. **Strength (S):** The evidence base was comprised of 4 studies. Of these 4 were considered to have a good sample size, and 4 had a clear metric for effectiveness. **Transparency (T):** 4 studies included were published and peer-reviewed, 0 were from the grey literature, and 0 were anecdotal. Of the studies included, 1 had a published methodology, and 4 justified their rationale.

2 Impact: Increased frequency/severity of storms (including wind, rain and wave action) increases foraging difficulty and/or mortality

Summary:

Invasive mammals are a major threat to many seabird populations, and as such there is a well-established literature on mammal exclusion, management and eradication detailing effective methods and case studies. However, there are more limited options when the mammalian predator in question is itself a conservation target, or is not easily managed. Nevertheless, for many situations there are several, well-researched, actions available that can benefit seabird populations effectively.

Intervention	Evidence of Effectiveness	R	S	T
Provide supplementary food during the breeding season	If storms affect foraging during the breeding season, it may be possible to support populations with additional food. Alternatively it may counteract the poor condition of adults after a harsh winter. Trialled on many seabird species. Limited evidence for effectiveness in auks, and all known studies are on *F. arctica*. Typically very labour intensive and difficult given the remote and inaccessible breeding colonies of auks. Likely only plausible for small populations.	3	4	3
Provide supplementary food during the non-breeding season	This is a hypothetical action. We found no published studies assessing this action's effectiveness for seabirds. It is likely to be very difficult or even impossible, especially for pelagic species.	NA	NA	NA
Rehabilitate sick or injured birds	For groups of long-lived, large birds, rehabilitation is known to be an effective way to support populations. However, examples in seabirds are scarce and the overall effectiveness for most species is unknown. Numerous rescue centres report successful rehabilitation of razorbills and murres, and some limited examples in puffins and little auks (survival rates unknown).	1	2	4

Green = Likely to be beneficial. Red = Unlikely to be beneficial, may have negative impact. Orange = contradicting or uncertain evidence. Grey = Limited evidence. R = relevance rating. S = strength rating. T = transparency rating. All ratings on a scale of 1 to 5, where 5 is the highest.

Details:

Provide supplementary food during the breeding season

Relevance (R): 5 studies in the evidence base focus on auks, 11 on other seabirds and 0 on other birds. **Strength (S):** The evidence base was comprised of 16 studies. Of these 10 were considered to have a good sample size, and 14 had a clear metric for effectiveness. **Transparency (T):** 16 studies included were published and peer-reviewed, 0 were from the grey literature, and 0 were anecdotal. Of the studies included, 13 had a published methodology, and 4 justified their rationale.

Rehabilitate sick or injured birds

Relevance (R): 0 studies in the evidence base focus on auks, 3 on other seabirds and 4 on other birds. **Strength (S):** The evidence base was comprised of 7 studies. Of these 4 were considered to have a good sample size, and 1 had a clear metric for effectiveness. **Transparency (T):** 7 studies included were published and peer-reviewed, 0 were from the grey literature, and 0 were anecdotal. Of the studies included, 5 had a published methodology, and 5 justified their rationale.

3 Impact: Increased frequency/severity of storms (including wind, rain and wave action) causes nest destruction

Summary:
While there are several local actions that may prevent or mitigate local nest destruction, they have not been trialled widely, and wide-spread evidence to support their use is currently lacking. If changes in extreme weather threatens the viability of a population, then several actions are available to encourage translocation of populations to safer areas.

Intervention	Evidence of Effectiveness	R	S	T
Alter habitat to encourage birds to leave an area	Few trials on seabirds and none on auks. Several trials of this action have been successful and encouraged terns to shift breeding sites. However, this action is likely more viable for species with lower site fidelity and areas with other available breeding habitat nearby.	2	2	3

Artificially incubate or hand-rear chicks to support population	Known to be effective for some seabirds, though labour intensive and usually only appropriate for small populations. Limited evidence in auks; ex-situ populations of puffins, murres and razorbills have been hand-reared and bred, though with low success rates.	2	2	3
Install barriers to prevent flooding	While likely to prevent flooding there is currently no evidence available on this action's effectiveness in relation to seabird conservation.	NA	NA	NA
Make new colonies more attractive to encourage birds to colonise	Several actions have been trialled across auk (and other seabird) species to encourage colonisation, with variable success, including the use of decoys, acoustic cues, smells and improved habitat. The most notable success has been to use decoys to encourage *F. arctica* to colonise new areas, other actions have had variable success depending on context and species.	2	4	3
Manually relocate nests	This has been reported by practitioners as an effective action to assist seabirds on low-lying beaches in the Baltic, including auks. However, to our knowledge there are no broad-scale studies or reviews of this action's effectiveness.	NA	NA	NA
Provide additional shelter or protection from extreme weather (flooding)	In some seabird species additional protection has reduced flooding, but evidence is limited. We found no published trials on auk species.	1	3	5
Provide artificial nesting sites	Tried extensively on many seabird species with notable success in many cases. Few trials for auk species, but some limited evidence for success in *F. arctica*, *S. antiquus* and *C. monocerata*.	3	5	3

		R	S	T
Repair/support nests to support breeding	Very limited evidence for effectiveness, but at least one case study has used this action to increase *U. aalge* breeding success.	3	5	3
Translocate the population to a more suitable breeding area	Known to be beneficial in other seabird groups, but evidence for auks is limited. At least one successful translocation of *F. arctica* has been carried out, but whether it is generally advisable is uncertain.	3	4	4

Green = Likely to be beneficial. Red = Unlikely to be beneficial, may have negative impact. Orange = contradicting or uncertain evidence. Grey = Limited evidence.
R = relevance rating. S = strength rating. T = transparency rating. All ratings on a scale of 1 to 5, where 5 is the highest.

Details:

Alter habitat to encourage birds to leave an area
Relevance (R): 0 studies in the evidence base focus on auks, 2 on other seabirds and 0 on other birds. **Strength (S):** The evidence base was comprised of 2 studies. Of these 2 were considered to have a good sample size, and 0 had a clear metric for effectiveness. **Transparency (T):** 2 studies included were published and peer-reviewed, 0 were from the grey literature, and 0 were anecdotal. Of the studies included, 2 had a published methodology, and 1 justified their rationale.

Artificially incubate or hand-rear chicks to support population
Relevance (R): 6 studies in the evidence base focus on auks, 34 on other seabirds and 0 on other birds. **Strength (S):** The evidence base was comprised of 40 studies. Of these 9 were considered to have a good sample size, and 19 had a clear metric for effectiveness. **Transparency (T):** 26 studies included were published and peer-reviewed, 0 were from the grey literature, and 0 were anecdotal. Of the studies included, 17 had a published methodology, and 4 justified their rationale.

Make new colonies more attractive to encourage birds to colonise
Relevance (R): 1 study in the evidence base focusses on auks, 37 on other seabirds and 6 on other birds. **Strength (S):** The evidence base was comprised of 44 studies. Of these 31 were considered to have a good sample size, and 18 had a clear metric for effectiveness. **Transparency (T):** 44 studies included were published and peer-reviewed, of which 1 were literature reviews or meta-analyses, 0 were from the grey literature, and 0 were anecdotal. Of the studies included, 30 had a published methodology, and 22 justified their rationale.

Provide additional shelter or protection from extreme weather (flooding)

Relevance (R): 0 studies in the evidence base focus on auks, 0 on other seabirds and 1 on other birds. **Strength (S):** The evidence base was comprised of 3 studies. Of these 1 was considered to have a good sample size, and 2 had a clear metric for effectiveness. **Transparency (T):** 3 studies included were published and peer-reviewed, 0 were from the grey literature, and 0 were anecdotal. Of the studies included, 3 had a published methodology, and 3 justified their rationale.

Provide artificial nesting sites

Relevance (R): 4 studies in the evidence base focus on auks, 48 on other seabirds and 1 on other birds. **Strength (S):** The evidence base was comprised of 54 studies. Of these 50 were considered to have a good sample size, and 33 had a clear metric for effectiveness. **Transparency (T):** 53 studies included were published and peer-reviewed, of which 2 were literature reviews or meta-analyses, 0 were from the grey literature, and 0 were anecdotal. Of the studies included, 33 had a published methodology, and 27 justified their rationale.

Repair/support nests to support breeding

Relevance (R): 1 study in the evidence base focusses on auks, 1 on other seabirds and 1 on other birds. **Strength (S):** The evidence base was comprised of 3 studies. Of these 1 was considered to have a good sample size, and 1 had a clear metric for effectiveness. **Transparency (T):** 3 studies included were published and peer-reviewed, 0 were from the grey literature, and 0 were anecdotal. Of the studies included, 1 had a published methodology, and 3 justified their rationale.

Translocate the population to a more suitable breeding area

Relevance (R): 1 study in the evidence base focusses on auks, 14 on other seabirds and 0 on other birds. **Strength (S):** The evidence base was comprised of 15 studies. Of these 13 were considered to have a good sample size, and 9 had a clear metric for effectiveness. **Transparency (T):** 14 studies included were published and peer-reviewed, of which 1 were literature reviews or meta-analyses, 0 were from the grey literature, and 0 were anecdotal. Of the studies included, 11 had a published methodology, and 9 justified their rationale.

4 Impact: Increased thermal stress

Summary:

There are currently no well-researched methods to directly assist seabirds with thermal stress, and more information is needed on how thermal stress can impact seabirds and how local conservation action can mitigate these impacts. If thermal stress becomes so common or extreme that it threatens the viability of a population, then several actions are available to encourage translocation of populations to safer areas.

Intervention	Evidence of Effectiveness	R	S	T
Make new colonies more attractive to encourage birds to colonise	Several actions have been trialled across auk (and other seabird) species to encourage colonisation, with variable success, including the use of decoys, acoustic cues, smells and improved habitat. The most notable success has been to use decoys to encourage *F. arctica* to colonise new areas, other actions have had variable success depending on context and species.	3	4	3
Provide additional resources to help seabirds thermoregulate (e.g. artificial pools)	This is a hypothetical action. We found no published studies assessing this action's effectiveness for seabirds.	NA	NA	NA
Provide additional shelter or protection from extreme weather (heatwaves)	Few trials on seabirds and none on auks. Additional shelter has been shown to protect cormorants from heatwaves, but more research is needed before this action can be generally recommended.	2	3	3

Intervention	Evidence of Effectiveness	R	S	T
Translocate the population to a more suitable breeding area	Known to be beneficial in other seabird groups, but evidence for auks is limited. At least one successful translocation of *F. arctica* has been carried out, but whether it is generally advisable is uncertain.	3	4	4

Green = Likely to be beneficial. Red = Unlikely to be beneficial, may have negative impact. Orange = contradicting or uncertain evidence. Grey = Limited evidence.
R = relevance rating. S = strength rating. T = transparency rating. All ratings on a scale of 1 to 5, where 5 is the highest.

Details:

Make new colonies more attractive to encourage birds to colonise
Relevance (R): 1 study in the evidence base focusses on auks, 37 on other seabirds and 6 on other birds. **Strength (S):** The evidence base was comprised of 44 studies. Of these 31 were considered to have a good sample size, and 18 had a clear metric for effectiveness. **Transparency (T):** 44 studies included were published and peer-reviewed, of which 1 were literature reviews or meta-analyses, 0 were from the grey literature, and 0 were anecdotal. Of the studies included, 30 had a published methodology, and 22 justified their rationale.

Provide additional shelter or protection from extreme weather (heatwaves)
Relevance (R): 0 studies in the evidence base focus on auks, 1 on other seabirds and 0 on other birds. **Strength (S):** The evidence base was comprised of 1 study. Of these 1 was considered to have a good sample size, and 1 had a clear metric for effectiveness. **Transparency (T):** 1 study included were published and peer-reviewed, 0 were from the grey literature, and 0 were anecdotal. Of the studies included, 0 had a published methodology, and 1 justified their rationale.

Translocate the population to a more suitable breeding area
Relevance (R): 1 study in the evidence base focusses on auks, 14 on other seabirds and 0 on other birds. Strength (S): The evidence base was comprised of 15 studies. Of these 13 were considered to have a good sample size, and 9 had a clear metric for effectiveness. **Transparency (T):** 14 studies included were published and peer-reviewed, of which 1 were literature reviews or meta-analyses, 0 were from the grey literature, and 0 were anecdotal. Of the studies included, 11 had a published methodology, and 9 justified their rationale.

5 Impact: Negative changes in vegetation

Summary:

While there are limited trials on seabirds, there are several concrete examples where local management has increased productivity even in relatively large breeding populations.

Intervention	Evidence of Effectiveness	R	S	T
Remove problematic vegetation	Removing vegetation has been shown to benefit several seabird species, but the amount of evidence is limited in auks. Removal of problematic vegetation has resulted in an increase in *F. arctica* breeding success at several sites in Scotland.	2	4	4

Green = Likely to be beneficial. Red = Unlikely to be beneficial, may have negative impact. Orange = contradicting or uncertain evidence. Grey = Limited evidence.
R = relevance rating. S = strength rating. T = transparency rating. All ratings on a scale of 1 to 5, where 5 is the highest.

Details:

Remove problematic vegetation

Relevance (R): 2 studies in the evidence base focus on auks, 9 on other seabirds and 5 on other birds. **Strength (S):** The evidence base was comprised of 16 studies. Of these 12 were considered to have a good sample size, and 9 had a clear metric for effectiveness. **Transparency (T):** 16 studies included were published and peer-reviewed, of which 1 were literature reviews or meta-analyses, 0 were from the grey literature, and 0 were anecdotal. Of the studies included, 13 had a published methodology, and 13 justified their rationale.

6 Impact: Reduced prey availability during breeding season

Summary:

Several local actions may assist breeding populations on a small scale, but direct intervention on a large scale is likely to be extremely difficult. General conservation actions to protect fish stocks and local marine areas may be the most effective method. If a population is likely to suffer major losses, even with conservation help, then translocations could be considered.

Intervention	Evidence of Effectiveness	R	S	T
Artificially incubate or hand-rear chicks to support population	Known to be effective for some seabirds, though labour intensive and usually only appropriate for small populations. Limited evidence in auks; ex-situ populations of puffins, murres and razorbills have been hand-reared and bred, though with low success rates.	2	2	3
Make new colonies more attractive to encourage birds to colonise	Several actions have been trialled across auk (and other seabird) species to encourage colonisation, with variable success, including the use of decoys, acoustic cues, smells and improved habitat. The most notable success has been to use decoys to encourage *F. arctica* to colonise new areas, other actions have had variable success depending on context and species.	2	4	3
Provide supplementary food during the breeding season	Trialled on many seabird species. Limited evidence for effectiveness in auks, and all known studies are on *F. arctica*. Typically very labour intensive and difficult given the remote and inaccessible breeding colonies of auks. Likely only plausible for small populations.	3	4	3
Translocate the population to a more suitable breeding area	Known to be beneficial in other seabird groups, but evidence for auks is limited. At least one successful translocation of *F. arctica* has been carried out, but whether it is generally advisable is uncertain.	3	4	4

Green = Likely to be beneficial. Red = Unlikely to be beneficial, may have negative impact. Orange = contradicting or uncertain evidence. Grey = Limited evidence.
R = relevance rating. S = strength rating. T = transparency rating. All ratings on a scale of 1 to 5, where 5 is the highest.

Details:

Artificially incubate or hand-rear chicks to support population

Relevance (R): 6 studies in the evidence base focus on auks, 34 on other seabirds and 0 on other birds. **Strength (S):** The evidence base was comprised of 40 studies. Of these 9 were considered to have a good sample size, and 19 had a clear metric for effectiveness. **Transparency (T):** 26 studies included were published and peer-reviewed, 0 were from the grey literature, and 0 were anecdotal. Of the studies included, 17 had a published methodology, and 4 justified their rationale.

Make new colonies more attractive to encourage birds to colonise

Relevance (R): 1 study in the evidence base focusses on auks, 37 on other seabirds and 6 on other birds. **Strength (S):** The evidence base was comprised of 44 studies. Of these 31 were considered to have a good sample size, and 18 had a clear metric for effectiveness. **Transparency (T):** 44 studies included were published and peer-reviewed, of which 1 were literature reviews or meta-analyses, 0 were from the grey literature, and 0 were anecdotal. Of the studies included, 30 had a published methodology, and 22 justified their rationale.

Provide supplementary food during the breeding season

Relevance (R): 5 studies in the evidence base focus on auks, 11 on other seabirds and 0 on other birds. **Strength (S):** The evidence base was comprised of 16 studies. Of these 10 were considered to have a good sample size, and 14 had a clear metric for effectiveness. **Transparency (T):** 16 studies included were published and peer-reviewed, 0 were from the grey literature, and 0 were anecdotal. Of the studies included, 13 had a published methodology, and 4 justified their rationale.

Translocate the population to a more suitable breeding area

Relevance (R): 1 study in the evidence base focusses on auks, 14 on other seabirds and 0 on other birds. **Strength (S):** The evidence base was comprised of 15 studies. Of these 13 were considered to have a good sample size, and 9 had a clear metric for effectiveness. **Transparency (T):** 14 studies included were published and peer-reviewed, of which 1 were literature reviews or meta-analyses, 0 were from the grey literature, and 0 were anecdotal. Of the studies included, 11 had a published methodology, and 9 justified their rationale.

7 Impact: Reduced prey availability during non-breeding season

Summary:

In pelagic species, local intervention to assist populations is likely to be difficult or impossible. General conservation actions to preserve fish stocks and protect marine areas are likely the most effective conservation actions available.

Intervention	Evidence of Effectiveness	R	S	T
Further protections at sea	Additional regulation to protect seabirds at sea can directly and indirectly benefit many seabird species, and limit the impact of climate change.	1	3	3

Green = Likely to be beneficial. Red = Unlikely to be beneficial, may have negative impact. Orange = contradicting or uncertain evidence. Grey = Limited evidence.
R = relevance rating. S = strength rating. T = transparency rating. All ratings on a scale of 1 to 5, where 5 is the highest.

Details:

Further protections at sea

Relevance (R): 0 studies in the evidence base focus on auks, 1 on other seabirds and 8 on other birds. **Strength (S):** The evidence base was comprised of 9 studies. Of these 7 were considered to have a good sample size, and 3 had a clear metric for effectiveness. **Transparency (T):** 9 studies included were published and peer-reviewed, of which 2 were literature reviews or meta-analyses, 0 were from the grey literature, and 0 were anecdotal. Of the studies included, 3 had a published methodology, and 6 justified their rationale.

© Seppo Häkkinen

Ducks and Phalaropes
(Anatidae and Scolopacidae)

An assessment of climate change vulnerability and potential conservation actions for ducks and phalaropes in the North-East Atlantic

UNIVERSITY OF CAMBRIDGE

ZSL Institute of Zoology

https://doi.org/10.11647/OBP.0343.02

1 Long-tailed Duck *(Clangula hyemalis)*

1.1 Evidence for exposure

1.1.1 Potential changes in breeding habitat suitability (by 2100):

🟧 Current breeding area that is likely to become less suitable (89% of current range).

🟨 Current breeding area that is likely to remain suitable (8%).

🟩 Current breeding area that is likely to become more suitable (3%).

1.1.2 Current impacts attributed to climate change:

① **Negative Impact:** Wintering populations in Europe have declined due to climate change-driven changes in predation in breeding areas outside of Europe.

② **Negative Impact:** Range expansion of red foxes following milder winters has led to predation of ducks much further north than previously, and may be threatening the viability of northern populations.

③ **Neutral Impact:** Competition with non-native gobies has caused long-tailed ducks to switch prey, though there has been no observed change in mortality or condition. Goby invasion may have been assisted by climate change, though currently this is speculative.

1.1.3 Predicted changes in key prey species:

No key prey assessment was carried out for this species.

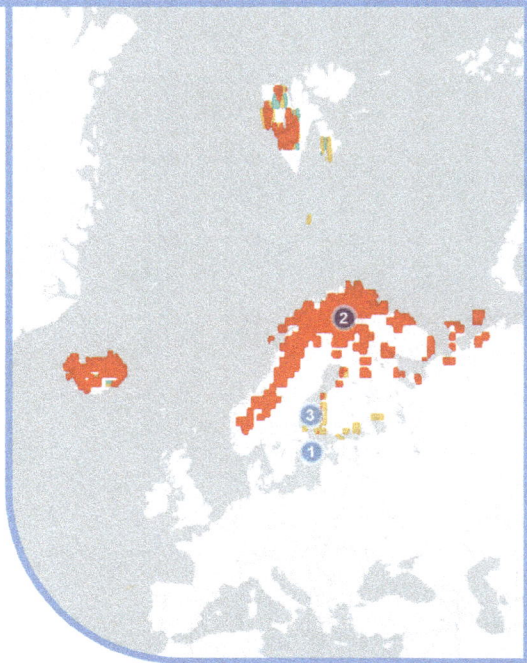

1.2 Sensitivity

• This species is numerous and has a large circumpolar range, but surveys suggest it is declining rapidly especially in the Baltic, most likely due to heavy bycatch. Any additional pressure from climate change is likely to exacerbate these declines.

• Several populations of long-tailed ducks are strongly reliant on *Mytilus edulis* for much of the year. *Mytilus* spp. are known to be sensitive to climate change, and warmer conditions are likely to result in lower quality prey. The consequences for long-tailed ducks are uncertain, but are very likely negative.

• Key wetland breeding habitats across the Arctic are rapidly disappearing or changing. The overall impact on long-tailed duck populations is unknown, but very likely to be negative.

• Species is known to be sensitive to changes in sea temperature, fluctuations in NAO, and, in particular, the presence of sea ice. A decline in sea ice is likely to result in a range shift of wintering populations, and it is possible that such a redistribution has already occurred in the Baltic.

• Long-tailed ducks often nest on low-lying areas near water, so are sensitive to flooding due to increased rainfall or wave-action. Any increase in intensity or frequency of storms is likely to impact breeding populations.

• Long-tailed ducks tend to winter in large groups in relatively small areas, so are vulnerable to mass mortality through extreme events. Even localised climate change impacts may have large consequences on the population as a whole.

1.3 Adaptive capacity

• Populations will mix during non-breeding season and there is low differentiation between populations. This may make populations more resilient to climate change, as immigration to support populations is more likely.

• Species has a very broad diet that varies depending on season and population. In the Baltic they have rapidly adjusted their diet in response to changes in prey availability. The loss of one prey species is unlikely to impact populations, though note that some populations may still be reliant on one or a few key species at some times of the year.

• Long-tailed ducks either abandon or skip breeding in particularly unsuitable years, preserving resources. This could be adaptive if conditions become more variable and ameliorate the impact of poor breeding conditions.

• Female long-tailed ducks generally have high site fidelity and are unlikely to shift breeding areas quickly in response to climate change.

2 Harlequin Duck (*Histrionicus histrionicus*)

1.1 Evidence for exposure

1.1.1 Potential changes in breeding habitat suitability (by 2100):

🟥 Current breeding area that is likely to become less suitable (95% of current range).

🟨 Current breeding area that is likely to remain suitable (2%).

🟩 Current breeding area that is likely to become more suitable (4%).

1.1.2 Current impacts attributed to climate change:

① **Neutral Impact:** Population has redistributed, with some populations growing and others shrinking, most likely due to shifts in prey species caused by climate change.

1.1.3 Predicted changes in key prey species:

No key prey assessment was carried out for this species.

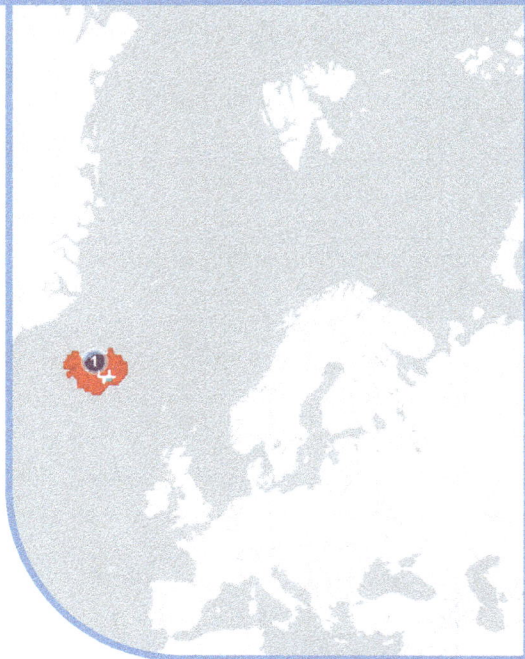

1.2 Sensitivity

• Harlequin ducks are sensitive to stream flow strength and variability during the breeding season; changes in freshwater flow can disrupt foraging and wash away nests. In parts of its range where temperature and precipitation patterns are likely to change, impacts could be significant.
• Harlequin ducks heavily rely on specific mollusc communities (e.g. *Mytilus* spp.), especially during the non-breeding season. Many marine mollusc species

are known to be sensitive to climate change, and warmer conditions are likely to result in reduced abundance of key prey species.

• Harlequin ducks tend to winter in large groups in relatively small areas, so are vulnerable to mass mortality through extreme events. Even localised climate change impacts may have large consequences on the population as a whole.

1.3 Adaptive capacity

• Fidelity to moulting and wintering locations is very high, less high to breeding sites. This makes the species vulnerable to changes in wintering sites, particularly to extreme events.

• The species changes breeding sites readily, as determined by local conditions. Redistributing in response to climate change could potentially buffer negative impacts.

© Seppo Häkkinen

3 Velvet Scoter *(Melanitta fusca)*

1.1 Evidence for exposure

1.1.1 Potential changes in breeding habitat suitability (by 2100):

■ Current breeding area that is likely to become less suitable (96% of current range).

■ Current breeding area that is likely to remain suitable (4%).

■ Current breeding area that is likely to become more suitable (0%).

1.1.2 Current impacts attributed to climate change:

① **Neutral Impact:** Scoters are starting their autumn migration significantly later in response to changing climate.

② **Neutral Impact:** Wintering populations have redistributed, most likely due to lack of prey caused at least partly by climate change.

1.1.3 Predicted changes in key prey species:

No key prey species are predicted to decline for this species.

1.1.4 Climate change impacts outside of Europe

• Climate change has contributed to declines of scoter populations in North America. Earlier spring snow melt has likely led to a trophic mismatch and lower breeding success in scoters.

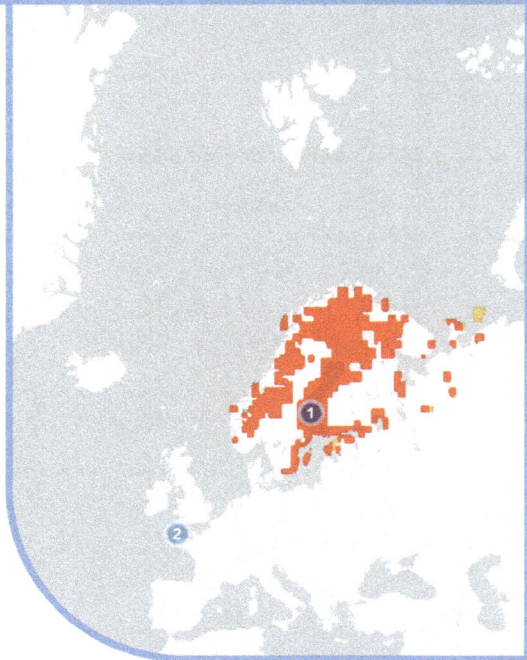

1.2 Sensitivity

• The species is numerous and has a large breeding range, but has suffered rapid declines since 1970. Particularly sensitive during non-breeding season as it gathers in high concentrations in relatively small areas (>90% of the global population winters in the Baltic). Additional pressure from climate change is likely to accelerate these declines.

• Many populations of velvet scoters rely on several key mollusc species for much of the year, many of which are known to be sensitive to climate change. Warmer conditions and ocean acidification are likely to result in reduced abundance of key prey species.

• Scoters, long tailed ducks and eiders have shown declines during historical regime shifts in marine ecosystems; they are likely sensitive to future changes in marine ecosystems.

• Key wetland breeding habitats across the Arctic are rapidly disappearing or changing. The overall impact on velvet scoters is unknown, but very likely to be negative.

• Often nests on low-lying areas near water, so is sensitive to flooding due to increased rainfall or wave-action. Any increase in intensity or frequency of storms is likely to impact breeding populations.

• Velvet scoters tend to winter in large groups in relatively small areas, so are vulnerable to mass mortality through extreme events. Even localised climate change impacts may have large consequences on the population as a whole.

• Recent observations and anecdotes have reported large groups of non-breeding scoters in the Baltic, many in poor condition. This may indicate high levels of stress or lack of high-quality prey. As this has not been observed previously, this may be a result of climate change. Further research is needed to clarify the cause of this and the long-term impact on the species.

1.3 Adaptive capacity

• Analyses of laying dates show there is little plasticity in response to changes in temperature, but there is some variation across and within populations.

• Species has shown flexibility in wintering sites in North America, in part predicted by year-to-year changes in sea temperature and fluctuations in NAO. Species is likely to shift wintering sites in response to climate change, and therefore buffer negative effects.

4 Common Scoter *(Melanitta nigra)*

1.1 Evidence for exposure

1.1.1 Potential changes in breeding habitat suitability (by 2100):

■ Current breeding area that is likely to become less suitable (94% of current range).

■ Current breeding area that is likely to remain suitable (5%).

■ Current breeding area that is likely to become more suitable (1%).

1.1.2 Current impacts attributed to climate change:

❶ **Neutral Impact:** Wintering populations have redistributed, most likely due to lack of prey caused at least partly by climate change.

1.1.3 Predicted changes in key prey species:

No key prey species are predicted to decline for this species.

1.1.4 Climate change impacts outside of Europe:

• Climate change has contributed to declines of scoter populations in North America. Earlier spring snow melt has likely led to a trophic mismatch and lower breeding success in scoters.

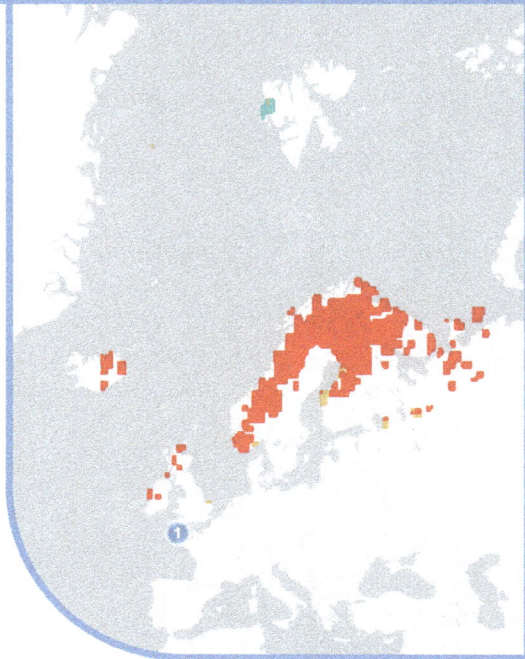

1.2 Sensitivity

• There is some evidence that populations in the Baltic have declined substantially, but these are unsubstantiated and may instead indicate population shifts to the North Sea. If the species is indeed declining, climate change is likely to exacerbate these declines.

• Many populations of common scoters rely on several key mollusc species for much of the year, many of which are known to be sensitive to climate change. Warmer conditions and ocean acidification are likely to result in reduced abundance of key prey species.

• Scoters have shown declines during historical regime shifts in marine ecosystems, they are likely sensitive to future changes in marine ecosystems.

• Key wetland breeding habitats across the Arctic are rapidly disappearing or changing. The overall impact on scoter populations is unknown, but is very likely to be significant.

• Often nests on low-lying areas near water, so is sensitive to flooding due to increased rainfall or wave-action. Any increase in intensity or frequency of storms is likely to impact breeding populations.

• There are records of common scoters in recent wrecks along the coast of the UK. The extent and severity of these wrecks is unknown, but may suggest that common scoters are vulnerable to extreme climate events, and may be significantly impacted by more frequent, or more extreme, storms.

• Recent observations and anecdotal reports have reported large groups of non-breeding scoters in the Baltic, many in poor condition. This may indicate high levels of stress or lack of high-quality prey. As this has not been observed previously, this may be a result of climate change. Further research is needed to clarify the cause of this and the long-term impact on the species.

1.3 Adaptive capacity

• If suitable habitat is available, common scoters have on occasion either temporarily or permanently colonised new areas. However, in general they have high site fidelity and are unlikely to shift breeding areas quickly in response to climate change.

• There is some evidence the wintering population in the Baltic is redistributing to the North Sea. However the extent of this shift is uncertain and the cause is unknown.

• Very varied diet which varies by population and by year. Most likely to be determined by prey abundance. The loss of one or a few prey species is unlikely to have a significant impact.

5 Red-breasted Merganser

(Mergus serrator)

1.1 Evidence for exposure

1.1.1 Potential changes in breeding habitat suitability (by 2100):

■ Current breeding area that is likely to become less suitable (92% of current range).

■ Current breeding area that is likely to remain suitable (7%).

■ Current breeding area that is likely to become more suitable (1%).

1.1.2 Current impacts attributed to climate change:

We did not identify any current impacts of climate change for this species.

1.1.3 Predicted changes in key prey species:

No key prey species are predicted to decline for this species.

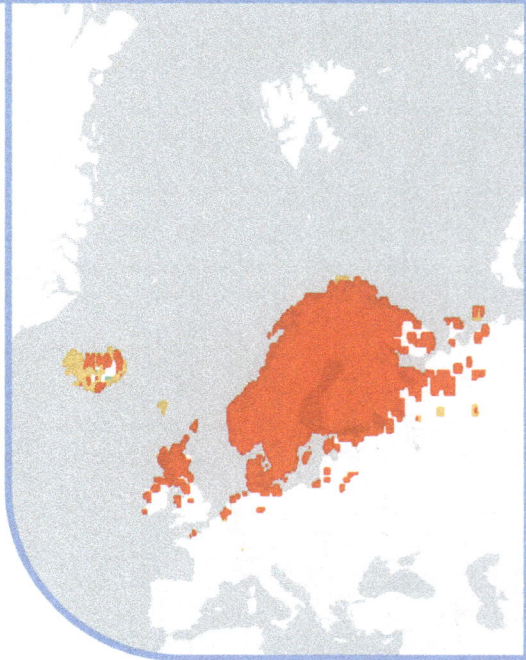

1.2 Sensitivity

• Red-breasted mergansers have low lying nests, which are vulnerable to flooding. Sea level rise or increased wave action from storms could impact breeding populations.

1.3 Adaptive capacity

• Red-breasted mergansers are capable of establishing or re-establishing colonies if conditions are suitable. Breeding range seems to be slowly expanding south in western Europe.

© Seppo Häkkinen

6 Red Phalarope *(Phalaropus fulicarius)*

1.1 Evidence for exposure

1.1.1 Potential changes in breeding habitat suitability (by 2100):

■ Current breeding area that is likely to become less suitable (93% of current range).

■ Current breeding area that is likely to remain suitable (6%).

■ Current breeding area that is likely to become more suitable (1%).

1.1.2 Current impacts attributed to climate change:

We did not identify any current impacts of climate change for this species.

1.1.3 Predicted changes in key prey species:

No key prey assessment was carried out for this species.

1.1.4 Climate change impacts outside of Europe:

• In Alaska red phalaropes now lay smaller eggs on average, presumably due to lower condition. This is likely due to delayed snow melt due to higher precipitation, despite the general warming trend.
• Across California red phalaropes have declined across their wintering areas. This is likely due to changes in ocean currents and declines in prey abundance.
• Populations around Alaska have declined in some areas, or possibly redistributed, due to changes in sea ice cover and in key copepod prey species.

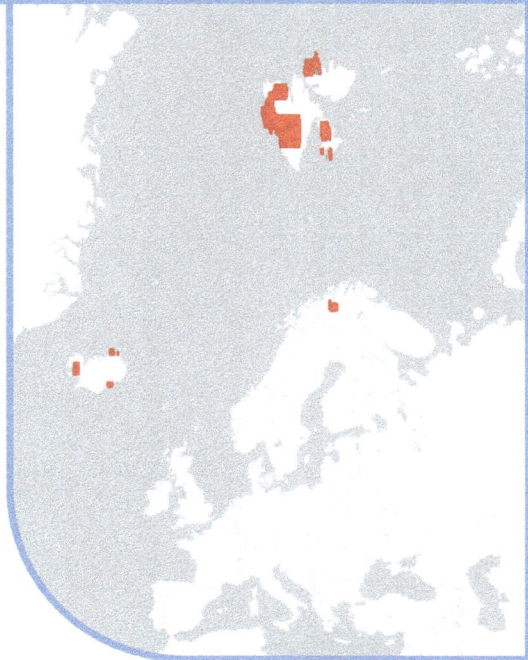

1.2 Sensitivity

• The species is sensitive to changes in Arctic tern (*Sterna paradisaea*) populations. It relies on predator alarm warning from breeding Arctic terns, and localised populations have decreased rapidly in some breeding colonies in Greenland in the absence of Arctic terns. Any climate change impacts on Arctic terns (which are documented) are likely to have an impact on phalaropes.

• The species is restricted to high-latitude wet tundra areas, which are predicted to considerably decrease in area over the next century. In some parts of its range, such habitat is already disappearing.

• Some recent mortalities in populations outside Europe have been linked to unusually warm weather, and this could be exacerbated by climate change.

• Population sizes, trends and threats are not well understood. Probably largest threats are climate change, predation by foxes and pollution. Any change in population size or impacts are not likely to be detected rapidly. Carrying out conservation action is likely to be challenging.

• Species is sensitive to many threats, including disturbance by forestry work, ship traffic, bycatch, wind farms and heavy tourism. Nest abandonment is common when disturbed. Conservation intervention may therefore be difficult.

• Phalaropes frequently gather in large groups in relatively small areas, so are vulnerable to mass mortality through extreme events. Even localised climate change impacts may have large consequences on the population as a whole.

1.3 Adaptive capacity

• Phalaropes in Alaska are known to change their laying date in response to changes in snow melt. No known study on phenology in Europe.

• Species has low site fidelity overall; red phalaropes readily change breeding sites depending on conditions. Species could potentially shift breeding sites in response to climate change.

• Red phalaropes either abandon or skip breeding in particularly poor years, preserving resources. This could be adaptive if conditions become more variable and ameliorate the impact of poor breeding conditions.

7 Red-necked Phalarope

(*Phalaropus lobatus*)

1.1 Evidence for exposure

1.1.1 Potential changes in breeding habitat suitability (by 2100):

■ Current breeding area that is likely to become less suitable (91% of current range).

■ Current breeding area that is likely to remain suitable (7%).

■ Current breeding area that is likely to become more suitable (2%).

1.1.2 Current impacts attributed to climate change:

1️⃣ **Neutral Impact:** Red-necked phalaropes have shifted north in Finland, the most southerly populations are declining while northerly populations are increasing. This shift is in correlation with climate change, but the underlying mechanism is not certain.

1.1.3 Predicted changes in key prey species:

No key prey assessment was carried out for this species.

1.1.4 Climate change impacts outside of Europe:

• A study in Alaska found that phalaropes have changed their laying date in response to changes in snow melt.

• Phalaropes have responded to changes in oceanic patterns in the Indian ocean and changed their foraging areas and patterns in response.

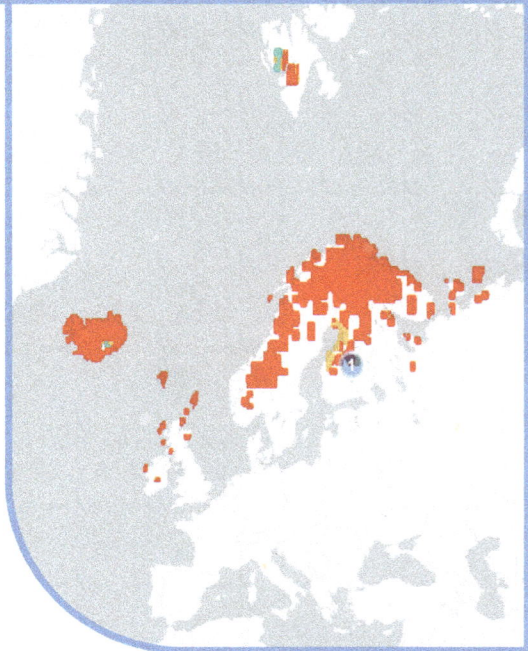

1.2 Sensitivity

• Very little is known about several aspects of this species' ecology. The status and trends of populations are largely unknown, and so any impacts of climate change will be difficult to identify, and effective conservation action will be challenging.

• Historical changes in weather patterns caused severe population crashes across America, Europe and Asia, many populations suffered enormous mortality and local near-extinctions. Impact was likely due to heavy depletion of plankton prey species.

• Phalaropes appear to have limited capacity to change prey species, and are heavily dependent on some key high-energy species during the breeding period. However, this is likely to vary depending on the population.

• Key wetland breeding habitats across the Arctic are rapidly disappearing or changing. The overall impact on phalaropes is unknown, but very likely to be negative.

• Phalaropes frequently gather in large groups in relatively small areas, so are vulnerable to mass mortality through extreme events. Even localised climate change impacts may have large consequences on the population as a whole.

• The species is sensitive to changes in Arctic tern (Sterna paradisaea) populations. It relies on predator alarm warning from breeding Arctic terns, and localised populations have decreased rapidly from some breeding colonies in Greenland in the absence of Arctic terns. Any climate change impacts on Arctic terns (which are documented) are likely to have an impact on phalaropes.

1.3 Adaptive capacity

• Phalaropes are known to change their laying date in response to environmental change, including recent changes in climate.

• Phalaropes undergo long migrations, which are energetically demanding. Different populations use different migration strategies and have adapted for different migration challenges. This may mean the species is flexible in its migration strategy and may respond to climate change, but also that the process is highly optimised. Changes in e.g. wind direction or strength could have large impacts.

• Red-necked phalaropes either abandon or skip breeding in particularly unsuitable years, preserving resources. This could be adaptive if conditions become more variable and ameliorate the impact of poor breeding conditions.

• Species has low site fidelity overall; red-necked phalaropes readily change breeding sites depending on conditions. Species could potentially shift breeding sites in response to climate change.

8 Steller's Eider *(Polysticta stelleri)*

1.1 Evidence for exposure

1.1.1 Potential changes in breeding habitat suitability (by 2100):

While this species does occasionally breed in Europe, recent assessments have concluded these populations are not permanent and are generally at very low densities. As such, no habitat suitability assessment was carried out.

1.1.2 Current impacts attributed to climate change:

Neutral Impact: Many Steller's eiders have changed wintering area from the Baltic to the White Sea, most likely due to decreases in sea ice. This may also be associated with an overall population decline, but this is uncertain.

1.1.3 Predicted changes in key prey species:

No key prey species are predicted to decline for this species.

1.2 Sensitivity

• This species has a large population and a large range, but is declining globally. In Europe the overall trend is unclear, there have been drastic declines in the Baltic, but this may be due to redistributions to other areas, in particular the north coast of Russia. Any existing declines are likely to be exacerbated by climate change.

- Eiders have shown declines during historical regime shifts in marine ecosystems, they are likely to be sensitive to future changes in marine ecosystems.
- Eiders are ground-nesters and vulnerable to predation by foxes and other mammals. Other eider species have suffered increased predation as a result of climate change. While this has not been observed in Steller's eiders, any increase in predation could have significant impacts on populations.
- Many of the species' key life-history traits are unknown. Many impacts to the population are unlikely to be detected quickly, and conservation action is likely to be challenging.
- Eiders form large rafts especially in the non-breeding season, making them particularly vulnerable to mass mortality events. Even localised climate change impacts may have large consequences on the population as a whole.
- Key wetland breeding habitats across the Arctic are rapidly disappearing or changing. The overall impact on eider populations is unknown, but very likely to be negative.
- Eiders have a varied diet of invertebrates but many populations are strongly reliant on *Mytilus* spp. and gastropods for much of the year. *Mytilus* spp. and gastropods are known to be sensitive to climate change, and warmer conditions are likely to result in reduction of key prey species. In addition, eiders show a preference for foraging in kelp beds, and any impacts to these may have severe negative consequences for foraging eiders.

1.3 Adaptive capacity

- While the species is generally considered to have high site fidelity to moulting and wintering sites, it has demonstrated range shifts in response to climate change. Large numbers have redistributed to northern Russia as sea ice has decreased over recent decades.
- Eiders either abandon or skip breeding in particularly unsuitable years, preserving resources. This could be adaptive if conditions become more variable and ameliorate the impact of poor breeding condition.

© Seppo Häkkinen

9 Common Eider *(Somateria mollissima)*

1.1 Evidence for exposure

1.1.1 Potential changes in breeding habitat suitability (by 2100):

■ Current breeding area that is likely to become less suitable (80% of current range).

■ Current breeding area that is likely to remain suitable (18%).

■ Current breeding area that is likely to become more suitable (2%).

1.1.2 Current impacts attributed to climate change:

① **Positive Impact:** Milder winter and summer weather have resulted in better average adult condition, and therefore better breeding success. In some areas this has resulted in local populations increases.

② **Neutral Impact:** Eiders have shifted their phenology in response to milder winters and lay earlier.

③ **Negative Impact:** Due to a lack of sea ice driven by climate change, polar bears are becoming more numerous around bird colonies during the summer and are more heavily predating on eider populations.

④ **Positive Impact:** Earlier melt of sea ice in spring has resulted in a decrease in predation by Arctic foxes, as they cannot access breeding colonies without the presence of sea ice.

⑤ **Positive Impact:** Earlier melt of sea ice in spring has resulted in an increase in eider population density, as eiders have earlier and longer access to high-quality prey.

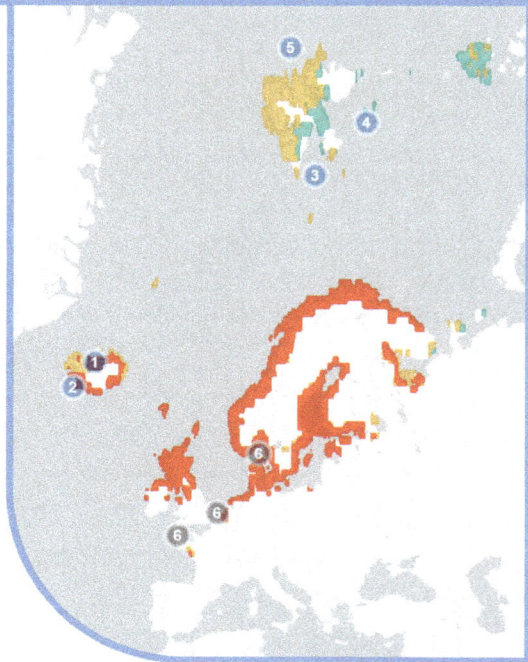

1.1.3 Predicted changes in key prey species:

⑥ Key prey species are likely to decline in abundance along the coast of Brittany, the coast of Belgium and along the south coast of Norway and Sweden.

1.1.3 Predicted changes in key prey species:

• Common eiders suffer increased predation from Arctic foxes due to prey switching following a collapse in lemming breeding cycles in northern Canada.

• In addition Canadian populations have suffered due to changes in weather in the breeding season, especially increased rain, either directly through exposure or indirectly through changes in predation.

1.2 Sensitivity

• This species has a large population and a large range, but is declining in several parts of its range. Known to be decreasing rapidly in the Baltic, climate change could be one of the underlying causes but this is uncertain.
• High precipitation and wind intensity reduce chick survival, likely through direct mortality from exposure and due to increased foraging difficulty. Precipitation is predicted to increase with climate change across northern Europe, and therefore likely to heavily affect chick survival.
• Often nests on low-lying coasts, so is sensitive to flooding due to increased rainfall or wave-action. Any increase in intensity or frequency of storms is likely to impact breeding populations.
• Eiders have a varied diet of invertebrates but many populations are strongly reliant on *Mytilus edulis* for much of the year. *Mytilus* spp. are known to be sensitive to climate change, and warmer conditions are likely to result in lower quality prey. The effect of this on eiders is uncertain, but is very likely negative.
• Eiders tend to winter in large groups in relatively small areas, so are vulnerable to mass mortality through extreme events. Even localised climate change impacts may have large consequences on the population as a whole.
• Eiders have shown declines during historical changes in sea temperature and fluctuations in weather patterns, particularly affecting the survival and distribution of wintering populations. They are likely sensitive to future changes in marine ecosystems.
• Eiders are particularly sensitive to disease outbreaks, bad years can claim up to 99% of offspring, and outbreaks seem to have a long term effect on female survival rate. Any change in disease dynamics due to climate change is likely to

have significant impacts.

• Warmer and calmer winters, with earlier break-up of sea ice, typically result in better condition adults and more adults committing to breeding in the following year. Many wintering areas (particularly in the Baltic and Svalbard) are likely to become milder due to climate change, which may benefit eider populations. However, sea ice also provides important resting areas for eiders during the winter, so a drastic reduction in sea ice may have counteracting negative effects.

• While climate change may have some positive impacts for common eiders, it does not mean that populations will necessarily increase in response to climate change. Recent research has found other pressures, in particular predation by native and non-native predators, may essentially override the positive effects of climate change, which appear to be small overall.

• This species has a long generation length (>10 years), which may slow recovery from severe impacts and increase population extinction risk.

1.3 Adaptive capacity

• Eiders either abandon or skip breeding in particularly unsuitable years, preserving resources. This could be adaptive if conditions become more variable and ameliorate the impact of poor breeding conditions.

• There is low migratory connectivity between many populations in Europe (though this is not universal), and many populations will mix in the non-breeding season. This could be adaptive, as low connectivity is associated with greater flexibility and higher responsiveness to change.

• Female common eiders are typically highly faithful to breeding sites; extreme breeding philopatry of female ducks means it is unlikely populations will shift in response to climate change.

• Historically eiders have expanded their range, and colonised new areas such as Ireland and NW England in the early 20th century. This suggests if populations grow and conditions are suitable then eiders can colonise new areas.

• While some populations appear to have changed their phenology, other studies have found changes in laying date are not related to weather conditions. Plasticity seems to vary between populations.

© Seppo Häkkinen

10 King Eider *(Somateria spectabilis)*

1.1 Evidence for exposure

1.1.1 Potential changes in breeding habitat suitability (by 2100):

🟥 Current breeding area that is likely to become less suitable (100% of current range).

🟨 Current breeding area that is likely to remain suitable (0%).

🟩 Current breeding area that is likely to become more suitable (0%).

1.1.2 Current impacts attributed to climate change:

We did not identify any current impacts of climate change for this species.

1.1.3 Predicted changes in key prey species:

① Key prey species are likely to decline in abundance on the west coast of Svalbard and around the Kanin Peninsula and the south Barents Sea.

1.1.4 Climate change impacts outside of Europe

• Increase in ice break-up, and increased variability of break-up, caused by climate change has resulted in significant damage to benthic prey and has caused local shifts in prey availability. Currently this has only a small impact on king eiders, but impacts could become significant in the future.

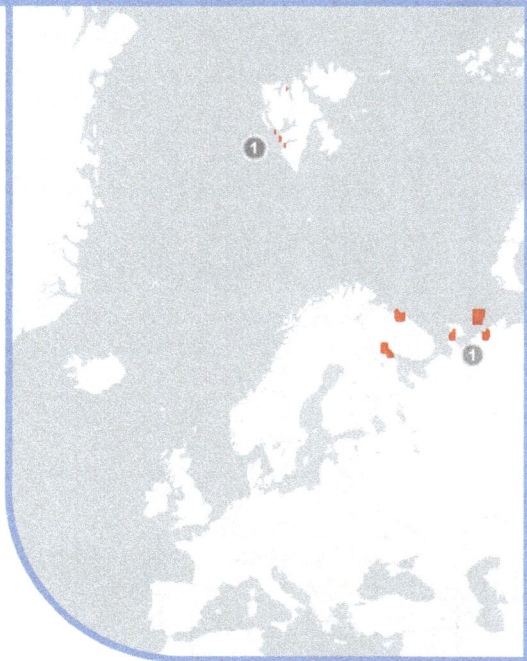

1.2 Sensitivity

• Eiders have shown declines during historical regime shifts in marine ecosystems, they are likely sensitive to future changes in marine regimes.

• Wetlands provide important breeding grounds for this species. Remote sensing and imaging has shown fluctuations in lakes in Siberia, with many lakes disappearing and others appearing. The impact on populations is unknown, as the area is difficult to study but the potential impact is very large.

• King eiders tend to winter in large groups in relatively small areas, so are vulnerable to mass mortality through extreme events. Even localised climate change impacts may have large consequences on the population as a whole.

• King eiders are vulnerable to mass mortality, particularly during migration. Hundreds of thousands of eiders have been recorded dying following unexpected re-freezing of sea ice. Change or variability of conditions during migration period could have significant impacts on the population.

• King eiders may be sensitive to the loss of sea ice, as it provides important roosting sites during winter and supports marine algae and benthic invertebrate prey. While impacts of the loss of sea ice have not been observed on eiders so far, it is a potential future impact.

• This species has a long generation length (>10 years), which may slow recovery from severe impacts and increases population extinction risk.

1.3 Adaptive capacity

• Very low site fidelity to breeding areas (especially in males), but also few documented examples of breeding in novel areas, and no documented records of permanent colonisation. It seems unlikely this species will shift their range rapidly in response to climate change. High site fidelity to wintering areas, and even local changes to these sites may have significant impact on populations.

• King eiders have a varied diet, and can dive to greater depths than many other marine ducks. They likely could switch to alternative prey or foraging strategies if climate change changes availability of prey.

• King eiders have varied migration pathways and strategies which vary between individuals and between years. This plasticity likely provides some resilience to climate change, as eiders could change their migration strategy in response to local conditions.

• King eiders have low migratory connectivity (populations will mingle during non-breeding season), and a weak genetic structure. This may be advantageous in response to climate change, as local adaptation is low and migration plasticity is high in king eiders which may allow them to respond rapidly to change.

• Eiders either abandon or skip breeding in particularly unsuitable years, preserving resources. This could be adaptive if conditions become more variable and ameliorate the impact of poor breeding conditions.

Potential actions in response to climate change: Ducks and Phalaropes (Anatidae and Scolopacidae)

In this section we list and assess possible local conservation actions that could be carried out in response to identified climate change impacts. This section is not grouped by species, but by identified impacts. If an impact or action is specific to one or a few species, this information is included in the action summary or in the footnotes. Effectiveness, relevance, strength and transparency scores are based on the available evidence we collated (see Appendix 2), and therefore all statements regarding limited or a lack of evidence relate to the collated evidence base, and does not infer that no such studies exist.

1 Impact: Increase in mammal predation

Summary:

Invasive mammals are a major threat to many seabird populations, and as such there is a well-established literature on mammal exclusion, management and eradication detailing effective methods and case studies. However, there are more limited options when the mammalian predator in question is itself a conservation target, or is not easily managed. Nevertheless, for many situations there are several, well-researched, actions available that can benefit seabird populations effectively.

Intervention	Evidence of Effectiveness	R	S	T
Manage/ eradicate mammalian predators	Strong evidence that predator management can assist seabird populations under heavy predation pressure, if carried out effectively. Various sea ducks have been shown to benefit from predator control, especially targeting rodents and mustelids. Larger predators, such as bears and foxes, are more difficult to deter, and other actions are more likely to be viable and/or effective.	2	5	3

		R	S	T
Physically protect nests with barriers or enclosures	Trialled extensively on many seabird groups, mostly with success, though depends on the species and the design of the barrier. Some trials on ducks have shown limited benefits, but generally quite minor and for limited species. Given the remote, dispersed nature of many duck and phalarope nesting sites, it may be difficult to carry this out at scale.	2	4	4
Reduce predation by translocating predators	Few trials on seabirds, and none for ducks and phalaropes. Existing evidence suggests this action can be beneficial and reduce egg/chick predation, and could be a possible action if other forms of predator management are not viable or permitted.	2	4	3
Repel predators with acoustic, chemical or visual deterrents	This is a hypothetical action. We found no records of this action's effectiveness for seabirds.	NA	NA	NA
Use supplementary feeding to reduce predation	Very few trials on seabirds, and none on ducks and phalaropes. No studies have shown this action is effective.	1	4	3

Green = Likely to be beneficial. Red = Unlikely to be beneficial, may have negative impact.
Orange = contradicting or uncertain evidence. Grey = Limited evidence.
R = relevance rating. S = strength rating. T = transparency rating. All ratings on a scale of 1 to 5, where 5 is the highest.

Details:

Manage/eradicate mammalian predators
Relevance (R): 1 study in the evidence base focusses on ducks and phalaropes, 45 on other seabirds and 3 on other birds. Strength (S): The evidence base was

comprised of 52 studies. Of these 44 were considered to have a good sample size, and 34 had a clear metric for effectiveness. Transparency (T): 52 studies included were published and peer-reviewed, of which 5 were literature reviews or meta-analyses, 0 were from the grey literature, and 0 were anecdotal. Of the studies included, 24 had a published methodology, and 28 justified their rationale.

Physically protect nests with barriers or enclosures
Relevance (R): 0 studies in the evidence base focus on ducks and phalaropes, 12 on other seabirds and 6 on other birds. Strength (S): The evidence base was comprised of 18 studies. Of these 16 were considered to have a good sample size, and 12 had a clear metric for effectiveness. Transparency (T): 17 studies included were published and peer-reviewed, 0 were from the grey literature, and 0 were anecdotal. Of the studies included, 11 had a published methodology, and 12 justified their rationale.

Reduce predation by translocating predators
Relevance (R): 0 studies in the evidence base focus on ducks and phalaropes, 2 on other seabirds and 2 on other birds. Strength (S): The evidence base was comprised of 4 studies. Of these 4 were considered to have a good sample size, and 3 had a clear metric for effectiveness. Transparency (T): 4 studies included were published and peer-reviewed, 0 were from the grey literature, and 0 were anecdotal. Of the studies included, 2 had a published methodology, and 3 justified their rationale.

Use supplementary feeding to reduce predation
Relevance (R): 0 studies in the evidence base focus on ducks and phalaropes, 1 on other seabirds and 3 on other birds. Strength (S): The evidence base was comprised of 4 studies. Of these 4 were considered to have a good sample size, and 4 had a clear metric for effectiveness. Transparency (T): 4 studies included were published and peer-reviewed, 0 were from the grey literature, and 0 were anecdotal. Of the studies included, 1 had a published methodology, and 4 justified their rationale.

© Seppo Häkkinen

© Seppo Häkkinen

© Image: Seppo Häkkinen

Gannets and Cormorants
(Sulidae and Phalacrocoracidae)

An assessment of climate change vulnerability and potential conservation actions for gulls and cormorants in the North-East Atlantic

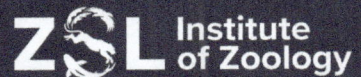

UNIVERSITY OF CAMBRIDGE

ZSL Institute of Zoology

https://doi.org/10.11647/OBP.0343.03

1 Northern Gannet *(Morus bassanus)*

1.1 Evidence for exposure

1.1.1 Potential changes in breeding habitat suitability (by 2100):

▇ Current breeding area that is likely to become less suitable (62% of current range).

▇ Current breeding area that is likely to remain suitable (37%).

▇ Current breeding area that is likely to become more suitable (1%).

1.1.2 Current impacts attributed to climate change:

① **Neutral Impact:** Gannets are undertaking longer foraging trips, most likely in response to prey shortages due to climate change. Although this likely increases the energetic costs of foraging, there have so far been no observed impacts on breeding success or mortality.

② **Positive Impact:** Gannets have established new colonies as key prey species have shifted further north.

1.1.3 Predicted changes in key prey species:

③ Key prey species are likely to decline in abundance in the southern Irish Sea and on the north coast of France.

1.1.4 Climate change impacts outside of Europe:

• Marine heatwaves in North America have resulted in wide-spread breeding failure and in some cases temporary desertion of colonies. Most likely because of prey shortages, but heat stress could play a role as well. It is difficult to

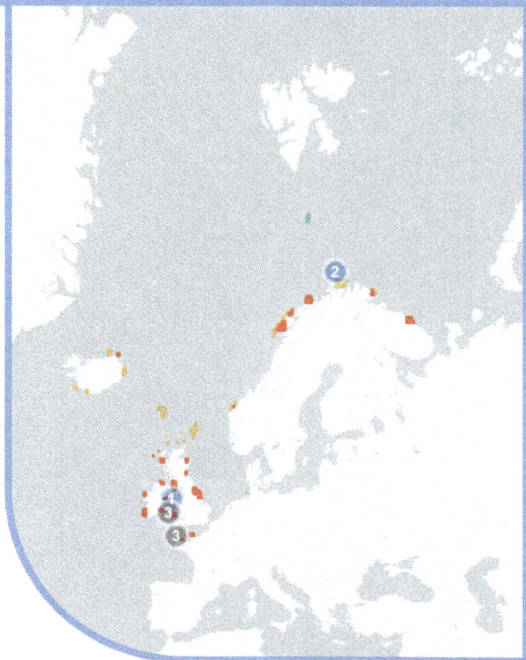

attribute individual climate events to climate change, but heatwaves are becoming more common and more extreme, and will likely continue to do so.

• Lack of key prey species (mackerel) due to warmer average marine temperatures and over-exploitation has caused low breeding success in a southern population of gannets in Canada.

1.2 Sensitivity

• Heatwaves are known to cause heat stress in gannet chicks and adults. So far this has not been observed to significantly affect populations in Europe, but heatwaves in other parts of their range have caused breeding failures and temporary colony desertion.
• This species has a long generation length (>10 years), which may slow recovery from severe impacts and increases population extinction risk.

1.3 Adaptive capacity

• Gannets occasionally establish new colonies, there are multiple records of them colonising or recolonising areas following environmental change or removal of threats.
• Individuals are often very loyal to breeding sites. Once adults establish a nest site they will return for many years, which reduces their capacity to adapt to change at breeding sites.
• Diet is variable across their range and over time. Long term studies have noticed shifts in primary prey species over several decades, which indicates some capacity for populations to shift diet.
• Gannets have been noted to change phenology, but not in correlation to changes in conditions. Overall there is little change in migration timing, and the underlying causes for any observed changes is uncertain.
• Gannets forage over large areas and show considerable flexibility in foraging behaviour, local changes in prey availability are unlikely to have a large impact.

Northern gannet © Silviu Petrovan

2 European Shag *(Gulosus aristotelis)*

1.1 Evidence for exposure

1.1.1 Potential changes in breeding habitat suitability (by 2100):

■ Current breeding area that is likely to become less suitable (44% of current range).

■ Current breeding area that is likely to remain suitable (60%).

■ Current breeding area that is likely to become more suitable (6%).

1.1.2 Current impacts attributed to climate change:

1 **Neutral Impact:** Shags have advanced their laying date, most likely due to changes in marine temperatures and subsequently in prey availability.

2 **Neutral Impact:** The diet composition of shags has changed a great deal, likely in response to climate change driven changes in the marine ecosystem.

3 **Negative Impact:** Extreme storms during the shag breeding season have led to wide-spread nest destruction, nesting failure and a net reduction in annual population production.

4 **Negative Impact:** Recent declines in shag populations because of high adult mortality are most likely because of increasingly severe winter storms.

5 **Neutral Impact:** Shags breed later as winters have become colder.

1.1.3 Predicted changes in key prey species:

6 Key prey species are likely to decline in abundance in the Irish Sea, the north-east of the UK, the northern coast of Spain and along the coast of Brittany

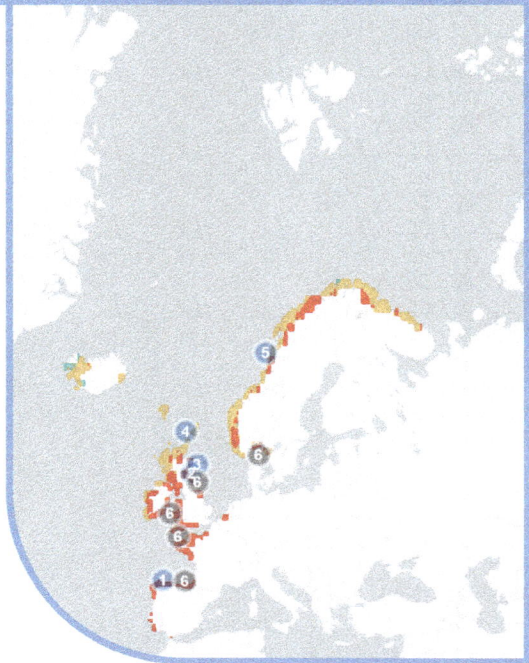

1.2 Sensitivity

• Shags only have partially water-proof feathers, and as such are prone to water-logging and hypothermia in wet, cold weather. Extreme events often result in wide-scale mortality ('wrecks') and breeding failure, as such predicted increases in storm frequency and intensity could have severe impacts on shags.

• When food is plentiful shag populations also often recover quite quickly, and so often follow a "boom and bust" cycle. This may be helpful in response to climate change as populations can recover quickly, but also makes them sensitive to rapid change as populations can become locally extinct quickly.

• While in many populations shags have a varied diet, in others they are heavily dependent on sandeels, saithe or herring. Previous decreases in prey species have led to lower breeding success. For some populations, any change to key prey availability due to climate change could have severe consequences.

• Shags typically nest on low-lying habitats, often only a few metres from the water-line, which are vulnerable to flooding and being washed away. Sea-level rise or an increase in wave action during the breeding season could have significant impacts on breeding colonies.

• Many shag populations heavily rely on kelp forest habitats, especially during the breeding season. Many kelp forests known to be important to shags are either declining due to climate change or are likely to be very vulnerable to climate change. So far, the impact of the decline in kelp forests on shags is unknown, but is potentially severe.

• Shags are likely to be sensitive to heatwaves, as they are commonly observed gular fluttering on hot days, and absorb heat very effectively. Other species of shags are known to be impacted by heatwaves, which cause poor breeding seasons and increased mortality. Heatwaves are likely to become more frequent and extreme due to climate change.

• The species is declining in many parts of its range. The causes behind these declines are not well understood, but climate change and extreme weather events are one likely cause.

1.3 Adaptive capacity

• Shags are known to change their phenology, probably in correlation to resource availability. This is likely to buffer some effects of climate change as shags may change breeding timing to match prey availability.

• Foraging strategy and behaviour varies substantially across populations and individuals. In general shags are flexible and generalist in terms of prey and foraging habitat, but individuals are often specialised.

3 Great Cormorant (*Phalacrocorax carbo*)

1.1 Evidence for exposure

1.1.1 Potential changes in breeding habitat suitability (by 2100):

🟧 Current breeding area that is likely to become less suitable (76% of current range).

🟨 Current breeding area that is likely to remain suitable (23%).

🟩 Current breeding area that is likely to become more suitable (1%).

1.1.2 Current impacts attributed to climate change:

① Neutral Impact: Cormorants that migrate to coastal areas during the winter are now migrating later, most likely due to less and later ice on freshwater feeding areas.

② Positive Impact: Cormorants are expanding their range due to increased availability of prey, in large part due to declines in competing marine predators, which in turn are partially driven by climate change.

1.1.3 Predicted changes in key prey species:

③ Key prey species are likely to decline in abundance in the Irish Sea, on the coast of Brittany, the north of Denmark, and across the English Channel.

1.1.4 Climate change impacts outside of Europe:

• Cormorants in Greenland have spread their summer range further north, most likely due to warmer sea temperatures and changes in food availability. However, this has also likely increased the costs of migration, as cormorants have further to travel to reach ice-free areas in winter.

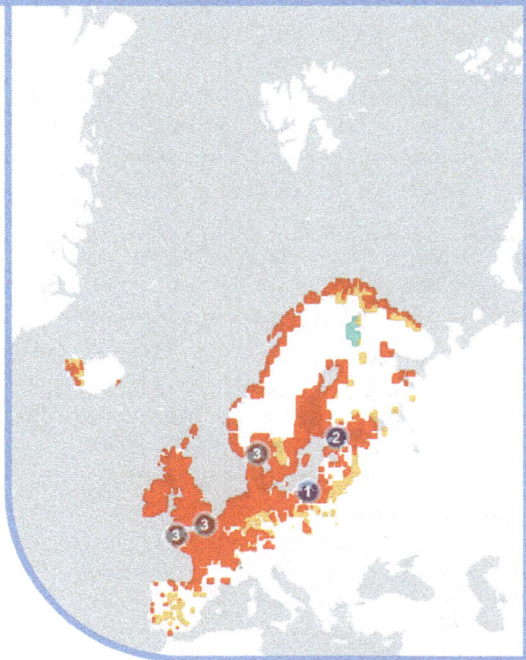

1.2 Sensitivity

• Cormorant population trends are correlated with local sea surface temperature, it seems likely that in many areas warmer waters will benefit cormorants. Sensitivity is therefore likely to be low in many areas. However, there is likely a limit to northern expansion as cormorants cannot forage effectively in areas with short periods of daylight.

• Cormorants are sensitive to extreme weather events, including extreme cold periods and high rainfall. Cormorants only have partially water-proof feathers, and as such are prone to water-logging and hypothermia in wet, cold weather. The northern edge of their range is likely set by winter temperatures, and the duration of sea and lake ice.

1.3 Adaptive capacity

• Following historical declines and several local extinctions, this species has greatly recovered following increased protection and decreased persecution, and has recolonised many areas of its previous range. It therefore seems likely that cormorants can redistribute and establish new populations in response to climate change.

• Cormorants generally have low site fidelity, and mass relocations are relatively common. It seems likely that cormorants can rapidly redistribute to more suitable areas if negative impacts occur.

• This species has a diverse diet, but this varies depending on population. Some populations variously prey on up to 20 species, depending on availability, whereas others are heavily on one or a few species. The effect of a loss of a key prey species is likely to vary depending on area and population.

European shag © Seppo Häkkinen

Potential actions in response to climate change: Gannets and Cormorants (Sulidae and Phalacrocoracidae)

In this section we list and assess possible local conservation actions that could be carried out in response to identified climate change impacts. This section is not grouped by species, but by identified impacts. If an impact or action is specific to one or a few species, this information is included in the action summary or in the footnotes. Effectiveness, relevance, strength and transparency scores are based on the available evidence we collated (see Appendix 2), and therefore all statements regarding limited or a lack of evidence relate to the collated evidence base, and does not infer that no such studies exist.

1 Impact: Increased frequency/severity of storms (including wind, rain and wave action) increases foraging difficulty and/or mortality

Summary:
Several local actions may be possible to limit mortality or increase recovery on a small scale, but for larger populations effective local action is difficult. Supporting the population in more general ways (increasing adult survival, limiting chick mortality) may be the most effective method.

Intervention	Evidence of Effectiveness	R	S	T
Provide supplementary food during the breeding season	Trialled on many seabird species. Limited evidence for effectiveness in gannets and cormorants, though it has been found to have some minor benefits for non-European gannets. Typically very labour intensive and difficult, and probably only plausible for small populations.	3	4	3

		R	S	T
Provide supplementary food during the non-breeding season	This is a hypothetical action. We found no published studies assessing this action's effectiveness for seabirds.	NA	NA	NA
Rehabilitate sick or injured birds	For groups of long-lived, large birds, rehabilitation is known to be an effective way to support populations. However, examples in seabirds are scarce and the overall effectiveness for most species is unknown. There are several reports of successful rehabilitation and release of gannets, cormorants and shags, but many note that treatment is difficult as these species are easily distressed, especially cormorants, and prone to disease in captivity.	1	2	3

Green = Likely to be beneficial. Red = Unlikely to be beneficial, may have negative impact.
Orange = contradicting or uncertain evidence. Grey = Limited evidence.
R = relevance rating. S = strength rating. T = transparency rating. All ratings on a scale of 1 to 5, where 5 is the highest.

Details:

Provide supplementary food during the breeding season
Relevance (R): 3 studies in the evidence base focus on gannets and cormorants, 13 on other seabirds and 0 on other birds. **Strength (S):** The evidence base was comprised of 16 studies. Of these 10 were considered to have a good sample size, and 14 had a clear metric for effectiveness. **Transparency (T):** 16 studies included were published and peer-reviewed, 0 were from the grey literature, and 0 were anecdotal. Of the studies included, 13 had a published methodology, and 4 justified their rationale.

Rehabilitate sick or injured birds
Relevance (R): 1 studies in the evidence base focus on gannets and cormorants, 2 on other seabirds and 4 on other birds. **Strength (S):** The evidence base was comprised of 7 studies. Of these 4 were considered to have a good sample size, and 1 had a clear metric for effectiveness. **Transparency (T):** 7 studies included were published and peer-reviewed, 0 were from the grey literature, and 0 were anecdotal. Of the studies included, 5 had a published methodology, and 5 justified their rationale.

2 Impact: Increased frequency/severity of storms (including wind, rain and wave action) causes nest destruction

Summary:

While there are several local actions that may prevent or mitigate local nest destruction, they have not been trialled widely and wide-spread evidence to support their use is currently lacking. If changes in extreme weather threatens the viability of a population, then several actions are available to encourage translocation of populations to safer areas.

Intervention	Evidence of Effectiveness	R	S	T
Alter habitat to encourage birds to leave an area	Few trials on seabirds and none on gannets or cormorants, has been effective in some other seabird groups. This action could be investigated for cormorants and shags, as this action is more viable for species lower site fidelity and areas with other available breeding habitat nearby.	2	2	3
Artificially incubate or hand-rear chicks to support population	Known to be effective for some seabirds, though labour intensive and usually only appropriate for small populations. No documented examples of hand-rearing in cormorants or gannets.	2	2	1
Install barriers to prevent flooding	While likely to prevent flooding there is currently no evidence available on this action's effectiveness in relation to seabird conservation.	NA	NA	NA
Make new colonies more attractive to encourage birds to colonise	Several actions have been trialled across seabird species to encourage colonisation, with variable success, including the use of decoys, acoustic cues, smells and improved habitat. There is limited evidence for cormorants and gannets, but several examples exist where cormorants have been successfully encouraged to new areas with artificial vocalisations and decoys.	2	4	3

Intervention	Evidence of Effectiveness	R	S	T
Manually relocate nests	This has been reported by practitioners as an effective action to assist seabirds on low-lying beaches in the Baltic, though on species other than cormorants and gannets. However, to our knowledge there are no broad-scale studies or reviews of this action's effectiveness.	NA	NA	NA
Provide additional shelter or protection from extreme weather (flooding)	There are few trials on seabird species, and most report little to no benefit for breeding populations. However, evidence is limited and more research is needed on this action's overall effectiveness. We found no published trials on gannet or cormorant species.	1	3	5
Provide artificial nesting sites	Tried extensively on many seabird species with significant benefit to many species. Artificial nesting sites have been successfully used to support several cormorant species; they readily visit, nest and breed at artificial nesting sites. No documented examples in gannets.	2	5	3
Repair/ support nests to support breeding	Very limited evidence for effectiveness in seabirds, though known to be effective in other birds. No known examples in gannet or cormorant species.	2	5	3
Translocate the population to a more suitable breeding area	Known to be beneficial in other seabird groups, but no documented examples in gannets or cormorants. Effectiveness and likelihood of success is poorly understood.	2	4	4

Green = Likely to be beneficial. Red = Unlikely to be beneficial, may have negative impact. Orange = contradicting or uncertain evidence. Grey = Limited evidence.
R = relevance rating. S = strength rating. T = transparency rating. All ratings on a scale of 1 to 5, where 5 is the highest.

Details:

Alter habitat to encourage birds to leave an area

Relevance (R): 0 studies in the evidence base focus on gannets and cormorants, 2 on other seabirds and 0 on other birds. **Strength (S):** The evidence base was comprised of 2 studies. Of these 2 were considered to have a good sample size, and 0 had a clear metric for effectiveness. **Transparency (T):** 2 studies included were published and peer-reviewed, 0 were from the grey literature, and 0 were anecdotal. Of the studies included, 2 had a published methodology, and 1 justified their rationale.

Artificially incubate or hand-rear chicks to support population

Relevance (R): 0 studies in the evidence base focus on gannets and cormorants, 40 on other seabirds and 0 on other birds. **Strength (S):** The evidence base was comprised of 40 studies. Of these 9 were considered to have a good sample size, and 19 had a clear metric for effectiveness. **Transparency (T):** 26 studies included were published and peer-reviewed, 0 were from the grey literature, and 0 were anecdotal. Of the studies included, 17 had a published methodology, and 4 justified their rationale.

Make new colonies more attractive to encourage birds to colonise

Relevance (R): 0 studies in the evidence base focus on gannets and cormorants, 38 on other seabirds and 6 on other birds. **Strength (S):** The evidence base was comprised of 44 studies. Of these 31 were considered to have a good sample size, and 18 had a clear metric for effectiveness. **Transparency (T):** 44 studies included were published and peer-reviewed, of which 1 were literature reviews or meta-analyses, 0 were from the grey literature, and 0 were anecdotal. Of the studies included, 30 had a published methodology, and 22 justified their rationale.

Provide additional shelter or protection from extreme weather (flooding)

Relevance (R): 0 studies in the evidence base focus on gannets and cormorants, 0 on other seabirds and 3 on other birds. **Strength (S):** The evidence base was comprised of 3 studies. Of these 1 was considered to have a good sample size, and 2 had a clear metric for effectiveness. **Transparency (T):** 3 studies included were published and peer-reviewed, 0 were from the grey literature, and 0 were anecdotal. Of the studies included, 3 had a published methodology, and 3 justified their rationale.

Provide artificial nesting sites

Relevance (R): 0 studies in the evidence base focus on gannets and cormorants, 53 on other seabirds and 1 on other birds. **Strength (S):** The evidence base was comprised of 54 studies. Of these 50 were considered to have a good sample size, and 33 had a clear metric for effectiveness. **Transparency (T):** 53 studies included were published and peer-reviewed, of which 2 were literature reviews or meta-analyses, 0 were from the grey literature, and 0 were anecdotal. Of the studies included, 33 had a published methodology, and 27 justified their rationale.

Repair/support nests to support breeding

Relevance (R): 0 studies in the evidence base focus on gannets and cormorants, 2 on other seabirds and 1 on other birds. **Strength (S):** The evidence base was comprised of 3 studies. Of these 1 was considered to have a good sample size, and 1 had a clear metric for effectiveness. **Transparency (T):** 3 studies included were published and peer-reviewed, 0 were from the grey literature, and 0 were anecdotal. Of the studies included, 1 had a published methodology, and 3 justified their rationale.

Translocate the population to a more suitable breeding area

Relevance (R): 0 studies in the evidence base focus on gannets and cormorants, 15 on other seabirds and 0 on other birds. **Strength (S):** The evidence base was comprised of 15 studies. Of these 13 were considered to have a good sample size, and 9 had a clear metric for effectiveness. **Transparency (T):** 14 studies included were published and peer-reviewed, of which 1 were literature reviews or meta-analyses, 0 were from the grey literature, and 0 were anecdotal. Of the studies included, 11 had a published methodology, and 9 justified their rationale.

© Seppo Häkkinen

Gulls

(Laridae)

An assessment of climate change vulnerability and potential conservation actions for gulls in the North-East Atlantic

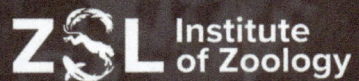

UNIVERSITY OF CAMBRIDGE

ZSL Institute of Zoology

https://doi.org/10.11647/OBP.0343.04

1 European Herring Gull

(Larus argentatus)

1.1 Evidence for exposure

1.1.1 Potential changes in breeding habitat suitability (by 2100):

■ Current breeding area that is likely to become less suitable (84% of current range).

■ Current breeding area that is likely to remain suitable (10%).

■ Current breeding area that is likely to become more suitable (5%).

1.1.2 Current impacts attributed to climate change:

🔵 **Negative Impact:** Changes in mercury cycling (due to increased sea temperatures) has led to increased exposure to mercury, with negative impacts on herring gull health.

1.1.3 Predicted changes in key prey species:

No key prey assessment was carried out for this species.

1.1.4 Climate change impacts outside of Europe

• Increased flooding due to sea level rise has led to the reduction or destruction of several populations in the US.

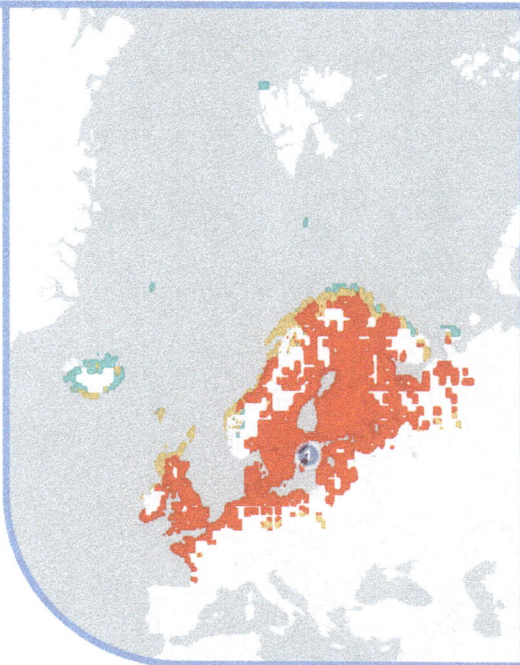

1.2 Sensitivity

• Herring gulls often nest in low-lying or exposed areas, which makes them vulnerable to storms and flooding. More frequent extreme storms or flooding during the breeding season could have severe effects on populations.
• During heatwaves, herring gulls' eggs and chicks have been observed suffering high stress and mortality. The overall impact of this on the population is unknown, but it appears herring gulls, especially in exposed areas, are vulnerable to heatwaves and increased frequency and intensity is likely to lower breeding success.
• This species has a long generation length (>10 years), which may slow recovery from severe impacts and increases population extinction risk.

1.3 Adaptive capacity

• Extremely variable diet, and able to exploit many available food sources. This is likely to make herring gulls more resilient to climate change, but note that many individual populations are specialised and are highly reliant on one or a few sources of food (e.g. human discards). Plasticity is therefore likely to vary across populations.
• Herring gulls are also adept at using urban environments which may buffer populations if natural diet or habitat is limited.
• Under the right circumstances, herring gulls can establish new colonies. While they tend to have some site fidelity (especially adults), they have been observed to colonise new areas over time if new areas are particularly high-quality or if previous area is disturbed.

2 Audouin's Gull *(Larus audouinii)*

1.1 Evidence for exposure

1.1.1 Potential changes in breeding habitat suitability (by 2100):

■ Current breeding area that is likely to become less suitable (68% of current range).

■ Current breeding area that is likely to remain suitable (32%).

■ Current breeding area that is likely to become more suitable (0%).

1.1.2 Current impacts attributed to climate change:

We did not identify any impacts of climate change on this species.

1.1.3 Predicted changes in key prey species:

① Key prey species are likely to decline in abundance on the south coast of Portugal.

1.2 Sensitivity

• The majority of the population is concentrated at relatively few breeding sites. This makes the European population as a whole vulnerable to change, including from climate change.

• Audouin's gulls are highly susceptible to other threats, in particular bycatch and predator disturbance. The species is currently sharply declining most likely due to changes in fishing discard practice and high predation rates. Any

additional pressure from climate change is likely to exacerbate these declines.
- The species is highly sensitive to changes in food availability. There can be rapid population growth in years of prey abundance, but rapid declines can occur in poor years. If climate change contributes to declines in key prey species, then gull populations are likely to be heavily impacted.
- Audouin's gulls frequently nest in low-lying or exposed areas (e.g. salt pans), which makes them vulnerable to storms and flooding. More frequent extreme storms or flooding during the breeding season could have severe effects on populations.

1.3 Adaptive capacity

- Following changes in fishing practice and other conservation measures, this previously endangered species recovered and expanded significantly. It appears to be able to grow in number and colonise new areas when conditions are suitable, indicating high dispersal ability.

© Silviu Petrovan

3 Caspian Gull *(Larus cachinnans)*

1.1 Evidence for exposure

1.1.1 Potential changes in breeding habitat suitability (by 2100):

■ Current breeding area that is likely to become less suitable (97% of current range).

■ Current breeding area that is likely to remain suitable (3%).

■ Current breeding area that is likely to become more suitable (0%).

1.1.2 Current impacts attributed to climate change:

We did not identify any current impacts of climate change for this species.

1.1.3 Predicted changes in key prey species:

No key prey assessment was carried out for this species.

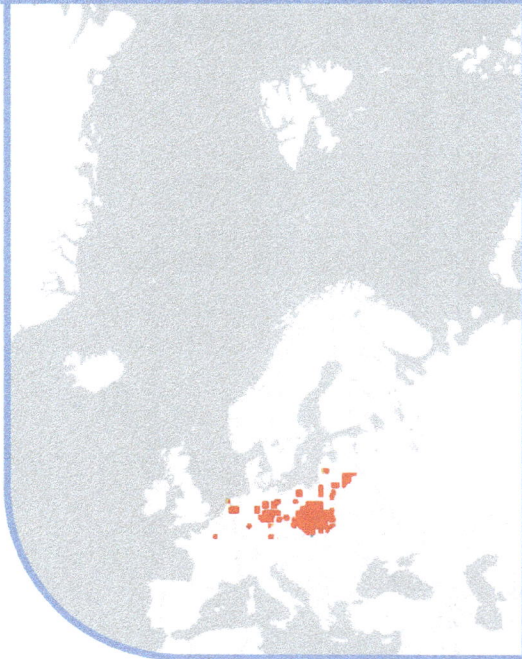

1.2 Sensitivity

• This species has a long generation length (>10 years), which may slow recovery from severe impacts and increases population extinction risk.

1.3 Adaptive capacity

• Caspian gulls have a varied diet and are likely able to prey switch. Change, or loss, of prey species due to climate change is unlikely to have wide-spread impact.

• Caspian gulls have recently expanded their range into several new areas of Europe and the species seems able to disperse and exploit suitable habitat effectively.

Herring gull chick © Silviu Petrovan

4 Lesser Black-backed Gull

(Larus fuscus)

1.1 Evidence for exposure

1.1.1 Potential changes in breeding habitat suitability (by 2100):

■ Current breeding area that is likely to become less suitable (61% of current range).

■ Current breeding area that is likely to remain suitable (27%).

■ Current breeding area that is likely to become more suitable (12%).

1.1.2 Current impacts attributed to climate change:

① **Positive Impact:** Positive Impact: Increased prey availability during the breeding season has led to population growth.

1.1.3 Predicted changes in key prey species:

② Key prey species are likely to decline in abundance on the south coast of Norway, the southern Irish Sea and along the Brittany Coast.

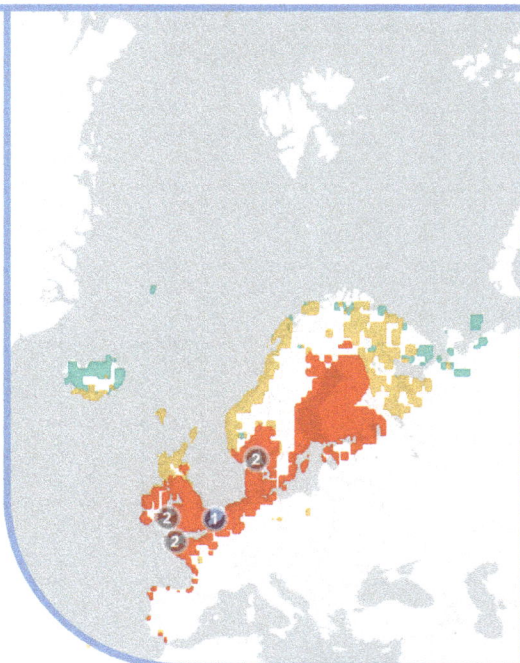

1.2 Sensitivity

• Lesser black-backed gulls typically nest in low-lying or exposed areas, which makes them vulnerable to storms and flooding. More frequent extreme storms or flooding during the breeding season could have severe effects on populations.

- During heatwaves, lesser black-backed gull eggs and chicks have been observed suffering high stress and mortality. The overall impact of this on the population is unknown, but it appears gulls, especially in exposed areas, are vulnerable to heatwaves and increased frequency and intensity is likely to lower breeding success.
- This species has a long generation length (>10 years), which may slow recovery from severe impacts and increases population extinction risk.

1.3 Adaptive capacity

- Extremely variable diet, and able to exploit many available food sources. This is likely to make lesser black-backed gulls more resilient to climate change, but note that many individual populations are specialised and are highly reliant on one or a few sources of food (e.g. human discards). In particular, the subspecies *Larus fuscus fuscus* in the Baltic are heavily reliant on spawning herring during the breeding season. Plasticity is therefore likely to vary across populations.
- Lesser black-backed gulls are also adept at using urban environments which may buffer populations if natural diet or habitat is limited.
- Under the right circumstances, lesser black-backed gulls can establish new colonies. While they tend to have some site fidelity (especially adults), they have been observed to colonise new areas over time if they are particularly high-quality or if previous area is disturbed.

5 Glaucous Gull (*Larus hyperboreus*)

1.1 Evidence for exposure

1.1.1 Potential changes in breeding habitat suitability (by 2100):

■ Current breeding area that is likely to become less suitable (77% of current range).

■ Current breeding area that is likely to remain suitable (23%).

■ Current breeding area that is likely to become more suitable (0%).

1.1.2 Current impacts attributed to climate change:

❶ Negative Impact: There has been increased predation by polar bears, most likely due to reduction in sea ice and therefore a lack of alternative prey. In some years this has severely affected breeding success.

❷ Negative Impact: Climate change is likely contributing to higher concentrations of contaminants ingested by glaucous gulls. The overall effect on the population is unknown, but presumably negative.

❸ Negative Impact: Climate change has contributed to a range shift in several helminth parasites, which has led to glaucous gulls being exposed to novel parasites, as well as increased parasite load. Effect on population is unknown, but presumably negative.

1.1.3 Predicted changes in key prey species:

❹ Key prey species are likely to decline in abundance along the Kanin Peninsula and southern Barents Sea.

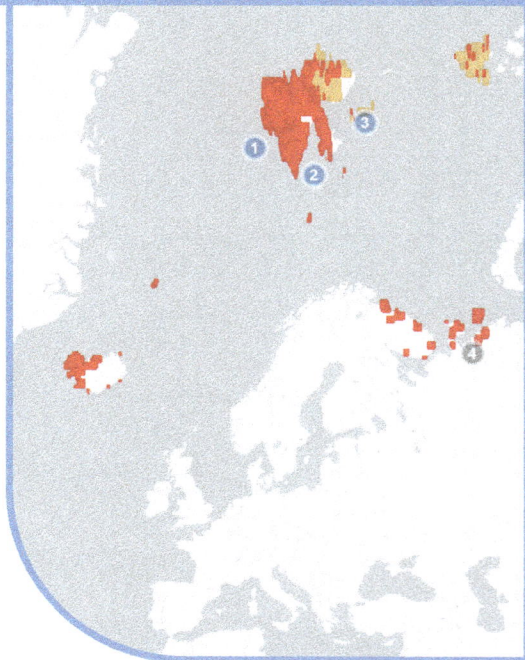

1.1.4 Climate change impacts outside of Europe

• Glaucous gull colonies display higher rates of cannibalism and lower breeding success in response to higher sea temperatures. This is presumably due to lack of marine prey, and is likely to be exacerbated with further climate change.

1.2 Sensitivity

• Chicks are susceptible to weather-related mortality, especially if severe wet weather occurs during hatching and first week post-hatch. Changes, especially an increase, in precipitation during key breeding periods may have large impacts on chick survival.

• Glaucous gulls have high accumulation of POP compounds, and current modelling suggests potential exacerbation of POPs and mercury in marine food webs due to climate change (i.e., increasing temperatures). Currently no negative impacts have been observed, but higher levels of bioaccummulation in the future is a potential risk to gull health.

• Avian flu has been recently recorded in some populations of glaucous gulls; warmer weather in the future may contribute to outbreaks.

• Competing species, such as herring gulls, are shifting their ranges north, in part in relation to climate change. This may lead to competition in the future if ranges overlap.

1.3 Adaptive capacity

• Very diverse diet, consisting of fish, marine invertebrates, bird eggs and young, small birds and mammals, carrion, refuse, seaweed, berries. Loss of one food source is unlikely to have a major impact on most populations.

6 Great Black-backed Gull
(Larus marinus)

1.1 Evidence for exposure

1.1.1 Potential changes in breeding habitat suitability (by 2100):

■ Current breeding area that is likely to become less suitable (87% of current range).

■ Current breeding area that is likely to remain suitable (26%).

■ Current breeding area that is likely to become more suitable (4%).

1.1.2 Current impacts attributed to climate change:

① **Negative Impact:** Higher sea temperatures correlate with lower breeding success. Mechanism unknown, but likely mediated through prey availability.

1.1.3 Predicted changes in key prey species:

② Key prey species are likely to decline in abundance in the Irish Sea, as well as along the Norwegian coast, southern coast of the UK and the Brittany Coast.

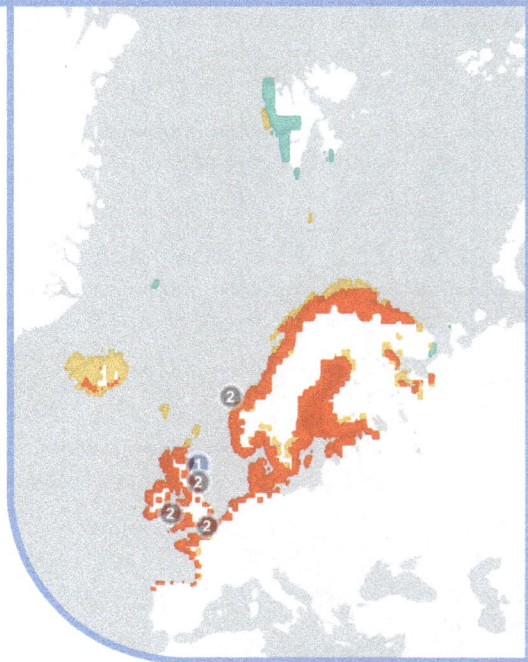

1.2 Sensitivity

• Great black-backed gulls are declining in areas of the eastern Atlantic, likely due to reductions in fishing discards. Any additional impacts, such as from climate change, are likely to exacerbate this decline.
• This species has a long generation length (>10 years), which may slow recovery from severe impacts and increases population extinction risk.

1.3 Adaptive capacity

• Extremely variable diet, and able to exploit many available food sources. This is likely to make great black-backed gulls more resilient to climate change, but note that many individual populations are specialised and are highly reliant on one or a few sources of food (e.g. human discards). Plasticity is therefore likely to vary across populations.

• Great black-backed gulls have historically shown range expansions when pressures have been alleviated, there is evidence they can colonise or re-colonise areas if they are particularly high-quality or if previous areas are disturbed.

• Great black-backed gulls occasionally use urban habitats and resources, which may buffer populations if natural diet or habitat is limited.

Herring gulls © Silviu Petrovan

7 Ivory Gull (*Pagophila eburnea*)

1.1 Evidence for exposure

1.1.1 Potential changes in breeding habitat suitability (by 2100):

■ Current breeding area that is likely to become less suitable (100% of current range).

■ Current breeding area that is likely to remain suitable (0%).

■ Current breeding area that is likely to become more suitable (0%).

1.1.2 Current impacts attributed to climate change:

❶ **Negative Impact:** Ivory gulls are heavily reliant on sea ice for breeding and hunting, recent decreases in sea ice are leading to rapid changes in population size and range.

❷ **Negative Impact:** As a secondary impact of sea ice loss, ivory gulls face more competition from other ivory gulls and from other species for resources.

1.1.3 Predicted changes in key prey species:

No key prey species are predicted to decline for this species.

1.1.4 Climate change impacts outside of Europe:

• Climate change is known to have several other impacts in other parts of the species range, in particular through changing winter food supplies, increasing competition with other marine birds, and increased predation due to increased access to previously isolated colonies.

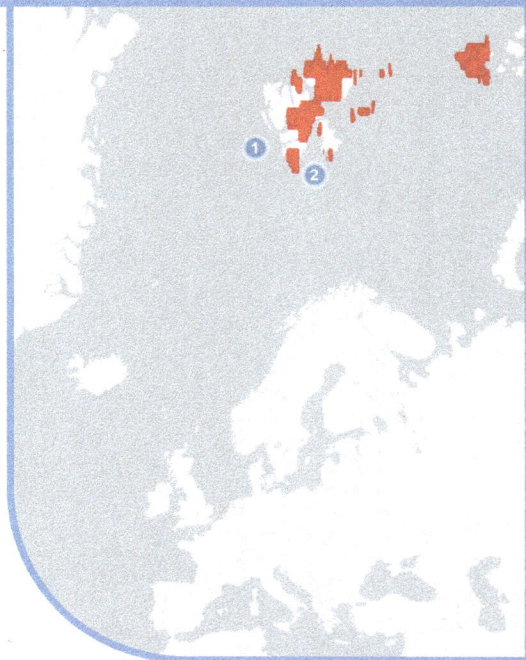

1.2 Sensitivity

•   Ivory gulls are highly dependent on sea ice; declines in population size and range have been linked in several areas to decreases in sea ice, particularly across the Canadian Arctic and Greenland.

•   Ivory gulls breed in extremely remote colonies which limits disturbance risk, but also makes monitoring and any potential conservation actions difficult.

•   Ivory gulls are sensitive to extreme climatic events; extreme heavy rainfall and windstorms have recently led to total breeding failures in Greenland. If climate change results in more extreme or more frequent extreme weather, this is likely to have severe impacts on ivory gull populations.

•   This species has a long generation length (>10 years), which may slow recovery from severe impacts and increases population extinction risk.

1.3 Adaptive capacity

•   Ivory gulls have a varied diet and are opportunistic feeders. The loss of one prey species is unlikely to have a major impact on the populations.

•   There is high connectivity and gene flow among populations, suggesting that populations are genetically diverse and there is significant exchange between populations. This could increase resilience to climate change as adaptive variation and immigration/emigration are more likely.

8 Black-legged Kittiwake

(Rissa tridactyla)

1.1 Evidence for exposure

1.1.1 Potential changes in breeding habitat suitability (by 2100):

🟧 Current breeding area that is likely to become less suitable (60% of current range).

🟨 Current breeding area that is likely to remain suitable (40%).

🟩 Current breeding area that is likely to become more suitable (0%).

1.1.2 Current impacts attributed to climate change:

① **Negative Impact:** Decreased prey availability due to warmer seas has led to lower breeding success.

② **Neutral Impact:** Kittiwake diet has changed significantly due to climate-change driven shift in prey assemblage. However, so far this has not resulted in any demonstrated change in breeding success.

③ **Neutral Impact:** Kittiwake populations have shifted their range in response to changes in distribution of key prey species.

④ **Neutral Impact:** Climate change has contributed to a range shift in several helminth parasites, which has led to kittiwakes being exposed to novel parasites, as well as increased parasite load. Effect on population is unknown, but most likely negative.

⑤ **Negative Impact:** Higher sea temperatures correlate with lower breeding success. Mechanism unknown, but potentially mediated through prey availability. Alternative theories suggest fishery pressure has been a large

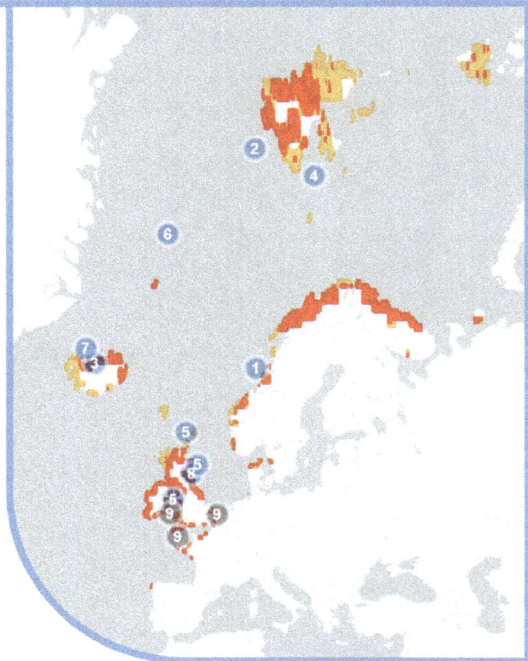

contributing factor.

(6) **Negative Impact:** Kittiwake colonies have declined during periods of rapid ocean warming. Mechanism unknown, but likely due to rapid changes in marine ecosystems and prey availability.

(7) **Negative Impact:** Extreme storms during the non-breeding season have led to mass mortality of kittiwakes ('wrecks').

(8) **Negative Impact:** Extreme storms during the kittiwake breeding season have led to wide-spread nest destruction, nesting failure and a net reduction in annual population production.

1.1.3 Predicted changes in key prey species:

(9) Key prey species are likely to decline in abundance in the Irish Sea, throughout the English Channel and along the Brittany Coast.

1.1.4 Climate change impacts outside of Europe
• Recent heatwaves in the North Pacific have resulted in mass mortality and wide-spread breeding failure at kittiwake colonies.

1.2 Sensitivity

• There appears to be strong variation in regional responses to climate change. The impacts of climate change on kittiwakes in Scotland have not been seen elsewhere in the UK. In addition, there is some debate on whether the drastic declines of kittiwake colonies in Scotland were primarily due to climate change or fisheries. The sensitivity of different populations to climate change is likely to vary.

• Large kittiwake colonies in the north Atlantic are supported indirectly by copepods (as they form the basis of the marine food chain). Projections of copepod abundance suggest they will range shift north, with large impacts on seabird colonies.

• Many kittiwake colonies are dependent on the timing of availability of key prey species, such as sandeels. Key prey species such as sandeels are known to be sensitive to warming temperatures, which may result in a phenological mismatch.

• Kittiwakes forage at or near the sea surface. If climate change results in more frequent or prolonged storms or prey moving into deeper water, it is likely to have significant impacts on kittiwake foraging.

1.3 Adaptive capacity

• There is some tentative evidence that kittiwakes can adaptively change their phenology based on studies in Svalbard. Populations in Scotland have also changed their laying date, possibly related to conditions in breeding and non-breeding areas.

• Under the right circumstances, kittiwakes can establish new colonies. While they tend to have some site fidelity (especially adults), they have been observed to colonise new areas over time if they are particularly high-quality or if previous area is disturbed.

• Kittiwakes occasionally use urban habitats and resources, especially where artificial nesting habitat is available, which may buffer populations if natural diet or habitat is limited.

© Silviu Petrovan

(This page and overleaf): Black-legged kittiwakes

9 Sabine's Gull *(Xema sabini)*

1.1 Evidence for exposure

1.1.1 Potential changes in breeding habitat suitability (by 2100):

■ Current breeding area that is likely to become less suitable (100% of current range).

■ Current breeding area that is likely to remain suitable (0%).

■ Current breeding area that is likely to become more suitable (0%).

1.1.2 Current impacts attributed to climate change:

We did not identify any current impacts of climate change for this species.

1.1.3 Predicted changes in key prey species:

No key prey assessment was carried out for this species.

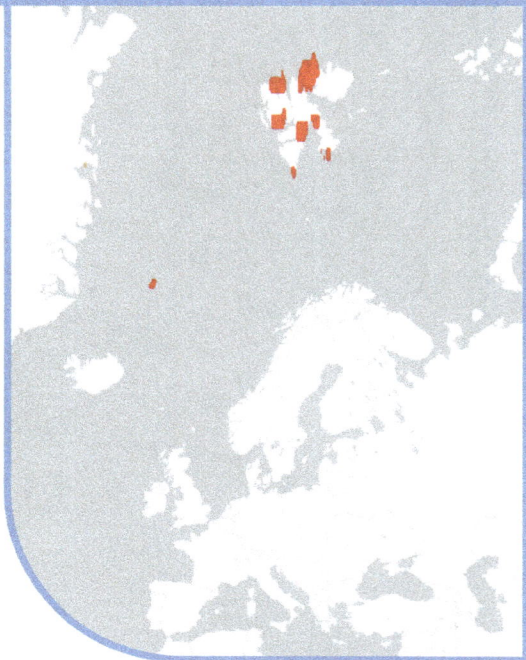

1.2 Sensitivity

•   As a high-Arctic species, it is likely that Sabine's gull is sensitive to climate change as the Arctic is currently undergoing rapid climate and ecological change. However no impacts have been observed so far.

•   There are generally few assessments of individual populations or their relative status, so impacts may not be recorded.

•   There is evidence that annual survival is affected by extreme climatic events in its tropical, non-breeding range; these may become more frequent with climate change.

• Sabine's gulls typically nest in low-lying, flooded areas, often very close to the high-tide line, which makes them vulnerable to storms and flooding. More frequent extreme storms or flooding during the breeding season could have severe effects on populations.

1.3 Adaptive capacity

• Populations of Sabine's gull overlap and meet during the non-breeding season (low migratory connectivity). Because of this, immigration is more likely to buffer climate change impacts, and higher genetic diversity of populations means adaptive response to climate change is more likely.

Black-legged kittiwake © Silviu Petrovan

10 Yellow-legged Gull

(Larus michahellis)

1.1 Evidence for exposure

1.1.1 Potential changes in breeding habitat suitability (by 2100):

■ Current breeding area that is likely to become less suitable (65% of current range).

■ Current breeding area that is likely to remain suitable (21%).

■ Current breeding area that is likely to become more suitable (14%).

1.1.2 Current impacts attributed to climate change:

① **Negative Impact:** Extreme storms during the razorbill breeding season have led to wide-spread nest destruction, nesting failure and a net reduction in annual population production.

1.1.3 Predicted changes in key prey species:

② Key prey species are likely to decline in abundance on the south coast of Portugal.

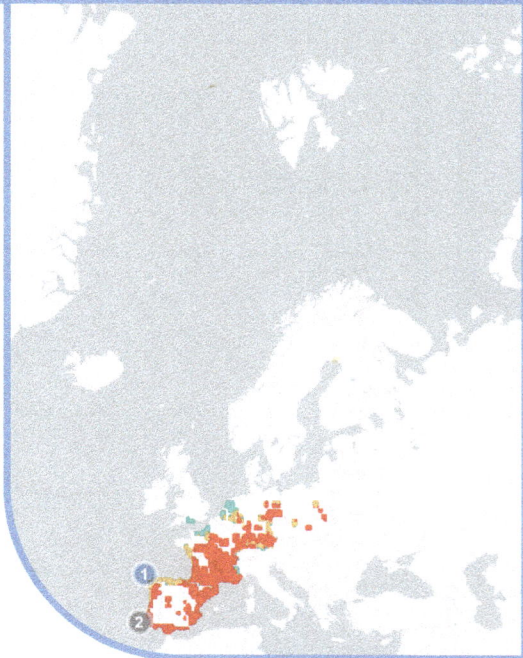

1.2 Sensitivity

• This species has a long generation length (>10 years), which may slow recovery from severe impacts and increases population extinction risk.

1.3 Adaptive capacity

• Very varied and opportunistic diet, including fish, invertebrates, mammals, refuse and offal, bird eggs and chicks. Yellow-legged gulls have been observed to change their primary prey species if one source becomes unavailable.

• Under the right circumstances, yellow-legged gulls can establish new colonies. While they tend to have some site fidelity (especially adults), they have been observed to colonise new areas over time if they are particularly high-quality or if previous area is disturbed.

Herring gulls © Silviu Petrovan

Potential actions in response to climate change: Gulls (Laridae)

In this section we list and assess possible local conservation actions that could be carried out in response to identified climate change impacts. This section is not grouped by species, but by identified impacts. If an impact or action is specific to one or a few species, this information is included in the action summary or in the footnotes. Effectiveness, relevance, strength and transparency scores are based on the available evidence we collated (see Appendix 2), and therefore all statements regarding limited or a lack of evidence relate to the collated evidence base, and does not infer that no such studies exist.

1 Impact: Increase in competition

Summary:
Local actions to prevent or mitigate the effects of competition are not well understood, and their effectiveness is unclear. In many contexts they are likely to be difficult or impossible to carry out on large populations. Supporting populations more generally (increasing adult survival, limiting chick mortality) may be a more appropriate strategy.

Intervention	Evidence of Effectiveness	R	S	T
Protect nest sites from competitors	Rarely trialled in seabirds, some benefits found in other non-seabird groups. Likely to be difficult due to large, cosmopolitan nature of many colonies; may be possible for species with spread-out, discrete nest-sites.	1	3	2
Reduce competition by removing competitor species	Trialled mostly on terns, but unclear if it is effective or not. Very scarce evidence for gulls, it has been trialled but the overall effectiveness is unclear. More research needed if this action is to be considered as a viable action.	3	3	3

| Use supplementary feeding to reduce competition | This is a hypothetical action. We found no published studies assessing this action's effectiveness for seabirds. | NA | NA | NA |

Green = Likely to be beneficial. Red = Unlikely to be beneficial, may have negative impact.
Orange = contradicting or uncertain evidence. Grey = Limited evidence.
R = relevance rating. S = strength rating. T = transparency rating. All ratings on a scale of 1 to 5, where 5 is the highest.

Details:

Protect nest sites from competitors
Relevance (R): 0 studies in the evidence base focus on gulls, 2 on other seabirds and 5 on other birds. **Strength (S):** The evidence base was comprised of 7 studies. Of these 5 were considered to have a good sample size, and 2 had a clear metric for effectiveness. **Transparency (T):** 6 studies included were published and peer-reviewed, 0 were from the grey literature, and 0 were anecdotal. Of the studies included, 3 had a published methodology, and 4 justified their rationale.

Reduce competition by removing competitor species
Relevance (R): 1 study in the evidence base focusses on gulls, 11 on other seabirds and 0 on other birds. **Strength (S):** The evidence base was comprised of 12 studies. Of these 10 were considered to have a good sample size, and 5 had a clear metric for effectiveness. **Transparency (T):** 12 studies included were published and peer-reviewed, 0 were from the grey literature, and 0 were anecdotal. Of the studies included, 8 had a published methodology, and 7 justified their rationale.

2 Impact: Increase in mammal predation

Summary:
Invasive mammals are a major threat to many seabird populations, and as such there is a well-established literature on mammal exclusion, management and eradication detailing effective methods and case studies. However, there are more limited options when the mammalian predator in question is itself a conservation target, or is not easily managed. Nevertheless, for many situations there are several, well-researched, actions available that can benefit seabird populations effectively.

Intervention	Evidence of Effectiveness	R	S	T
Manage/ eradicate mammalian predators	Strong evidence that predator management can assist seabird populations if under heavy predation pressure, and if carried out effectively. Several successful examples in gulls.	3	5	3
Physically protect nests with barriers or enclosures	Trialled extensively on many seabird groups, mostly with success, though depends on the species and the design of the barrier. Some trials on gulls, in particular on Audouin's gull, have shown benefits and lowered predation.	3	4	4
Reduce predation by translocating predators	Few trials on seabirds, and none for gulls. Existing evidence suggests this action can be beneficial and reduce egg/chick predation, and could be a possible action if other forms of predator management are not viable.	2	4	3
Repel predators with acoustic, chemical or visual deterrents	This is a hypothetical action. We found no published studies assessing this action's effectiveness for seabirds.	NA	NA	NA
Use supplementary feeding to reduce predation	Very few trials on seabirds, and none on gulls. No studies have shown this action is effective.	NA	NA	NA

Green = Likely to be beneficial. Red = Unlikely to be beneficial, may have negative impact.
Orange = contradicting or uncertain evidence. Grey = Limited evidence.
R = relevance rating. S = strength rating. T = transparency rating. All ratings on a scale of 1 to 5, where 5 is the highest.

Details:

Manage/eradicate mammalian predators
Relevance (R): 2 studies in the evidence base focus on gulls, 43 on other seabirds and 4 on other birds. **Strength (S):** The evidence base was comprised of 52 studies. Of these 44 were considered to have a good sample size, and 34 had a clear metric for effectiveness. **Transparency (T):** 52 studies included were published and peer-reviewed, of which 5 were literature reviews or meta-analyses, 0 were from the grey literature, and 0 were anecdotal. Of the studies included, 24 had a published methodology, and 28 justified their rationale.

Physically protect nests with barriers or enclosures
Relevance (R): 3 studies in the evidence base focus on gulls, 9 on other seabirds and 6 on other birds. **Strength (S):** The evidence base was comprised of 18 studies. Of these 16 were considered to have a good sample size, and 12 had a clear metric for effectiveness. **Transparency (T):** 17 studies included were published and peer-reviewed, 0 were from the grey literature, and 0 were anecdotal. Of the studies included, 11 had a published methodology, and 12 justified their rationale.

Reduce predation by translocating predators
Relevance (R): 0 studies in the evidence base focus on gulls, 2 on other seabirds and 2 on other birds. **Strength (S):** The evidence base was comprised of 4 studies. Of these 4 were considered to have a good sample size, and 3 had a clear metric for effectiveness. **Transparency (T):** 4 studies included were published and peer-reviewed, 0 were from the grey literature, and 0 were anecdotal. Of the studies included, 2 had a published methodology, and 3 justified their rationale.

Use supplementary feeding to reduce predation
Relevance (R): 0 studies in the evidence base focus on gulls, 1 on other seabirds and 3 on other birds. **Strength (S):** The evidence base was comprised of 4 studies. Of these 4 were considered to have a good sample size, and 4 had a clear metric for effectiveness. **Transparency (T):** 4 studies included were published and peer-reviewed, 0 were from the grey literature, and 0 were anecdotal. Of the studies included, 1 had a published methodology, and 4 justified their rationale.

3 Impact: Increased exposure to pollution and heavy metals

Summary:

The effects of pollution and heavy metals are known to have serious consequences for gulls, but despite this there are no current actions that are well-researched. It is likely prevention is more effective than treatment, so the most effective action in many cases is to deter gulls (if possible) from using a heavily polluted area.

Intervention	Evidence of Effectiveness	R	S	T
Alter habitat to encourage birds to leave an area	Very limited evidence for seabirds, and none for gulls. Several successful examples of this action in terns, but more research needed before this is considered as a viable option for gulls.	2	2	3
Reduce exposure to pollutants	This is a hypothetical action. We found no published studies assessing this action's effectiveness for seabirds.	NA	NA	NA
Treat sick or injured birds affected by pollution/ heavy metals	This is a hypothetical action. We found no published studies assessing this action's effectiveness for seabirds (and did not include evidence regarding treatment following oil spills). Likely to be resource intensive.	NA	NA	NA

Green = Likely to be beneficial. Red = Unlikely to be beneficial, may have negative impact. Orange = contradicting or uncertain evidence. Grey = Limited evidence.
R = relevance rating. **S** = strength rating. **T** = transparency rating. All ratings on a scale of 1 to 5, where 5 is the highest.

Details:

Alter habitat to encourage birds to leave an area

Relevance (R): 0 studies in the evidence base focus on gulls, 2 on other seabirds and 0 on other birds. **Strength (S):** The evidence base was comprised of 2 studies. Of these 2 were considered to have a good sample size, and 0 had a clear metric for effectiveness. **Transparency (T):** 2 studies included were published and peer-reviewed, 0 were from the grey literature, and 0 were anecdotal. Of the studies included, 2 had a published methodology, and 1 justified their rationale.

4 Impact: Reduced area of breeding or foraging habitat

Summary:
On a local scale, providing artificial nesting sites can be an effective method of counteracting this impact, though there are relatively few trials on gulls. Outside of this, if lack of habitat threatens the viability of a population, then several actions are available to encourage translocation of populations to safer areas.

Intervention	Evidence of Effectiveness	R	S	T
Alter habitat to encourage birds to leave an area	Few trials on seabirds and none on gulls. Several trials of this action have been successful and encouraged terns to shift breeding sites. However, this action is likely more viable for species with lower site fidelity and areas with other available breeding habitat nearby.	2	2	3
Make new colonies more attractive to encourage birds to colonise	Several different methods have been trialled extensively across other seabirds, with variable success depending on method and species. No evidence currently available for gulls, the effectiveness of decoys, acoustic cues, smells and improved habitat is currently unknown.	2	4	3
Provide artificial nesting sites	Tried extensively on many seabird species with significant benefit to many species. Artificial nesting sites have been successfully used to support kittiwake populations, for other gull species results have been mixed but several species have benefited from artificial nesting sites.	3	5	3
Translocate the population to a more suitable breeding area	Known to be beneficial in other seabird groups, but evidence for gulls is limited. Several failed attempts have been recorded, and to our knowledge no successful translocations of gulls have been carried out.	3	4	4

Green = Likely to be beneficial. Red = Unlikely to be beneficial, may have negative impact. Orange = contradicting or uncertain evidence. Grey = Limited evidence.
R = relevance rating. **S** = strength rating. **T** = transparency rating. All ratings on a scale of 1 to 5, where 5 is the highest.

Details:

Alter habitat to encourage birds to leave an area
Relevance (R): 0 studies in the evidence base focus on gulls, 2 on other seabirds and 0 on other birds. **Strength (S):** The evidence base was comprised of 2 studies. Of these 2 were considered to have a good sample size, and 0 had a clear metric for effectiveness. **Transparency (T):** 2 studies included were published and peer-reviewed, 0 were from the grey literature, and 0 were anecdotal. Of the studies included, 2 had a published methodology, and 1 justified their rationale.

Make new colonies more attractive to encourage birds to colonise
Relevance (R): 0 studies in the evidence base focus on gulls, 38 on other seabirds and 6 on other birds. **Strength (S):** The evidence base was comprised of 44 studies. Of these 31 were considered to have a good sample size, and 18 had a clear metric for effectiveness. **Transparency (T):** 44 studies included were published and peer-reviewed, of which 1 were literature reviews or meta-analyses, 0 were from the grey literature, and 0 were anecdotal. Of the studies included, 30 had a published methodology, and 22 justified their rationale.

Provide artificial nesting sites
Relevance (R): 1 study in the evidence base focusses on gulls, 51 on other seabirds and 1 on other birds. **Strength (S):** The evidence base was comprised of 54 studies. Of these 50 were considered to have a good sample size, and 33 had a clear metric for effectiveness. **Transparency (T):** 53 studies included were published and peer-reviewed, of which 2 were literature reviews or meta-analyses, 0 were from the grey literature, and 0 were anecdotal. Of the studies included, 33 had a published methodology, and 27 justified their rationale.

Translocate the population to a more suitable breeding area
Relevance (R): 1 study in the evidence base focusses on gulls, 14 on other seabirds and 0 on other birds. **Strength (S):** The evidence base was comprised of 15 studies. Of these 13 were considered to have a good sample size, and 9 had a clear metric for effectiveness. **Transparency (T):** 14 studies included were published and peer-reviewed, of which 1 were literature reviews or meta-analyses, 0 were from the grey literature, and 0 were anecdotal. Of the studies included, 11 had a published methodology, and 9 justified their rationale.

5 Impact: Reduced prey availability during breeding season

Summary:
Several local actions may assist breeding populations on a small scale, but direct intervention on a large scale is likely to be extremely difficult. General conservation actions to protect fish stocks and local marine areas may be the most effective method. If a population is likely to suffer major losses, even with conservation help, then translocations could be considered.

Intervention	Evidence of Effectiveness	R	S	T
Artificially incubate or hand-rear chicks to support population	Known to be effective for some seabirds, though labour intensive and usually only appropriate for small populations. Many gull species have been successfully hand-reared and bred, but typically in small numbers. Likely to be difficult for many species, especially those that breed in coastal, inaccessible habitats.	3	2	1
Make new colonies more attractive to encourage birds to colonise	Several different methods have been trialled extensively across other seabirds, with variable success depending on method and species. No evidence currently available for gulls, the effectiveness of decoys, acoustic cues, smells and improved habitat is currently unknown.	2	4	3
Provide supplementary food during the breeding season	Trialled on many seabird species. Known to be beneficial for several gull species, but success varies. Many gulls will scavenge any available food source, so it is feasible to provide supplementary food. However, as many gull populations are already reliant on discards, there are ethical concerns regarding wide-spread use of supplemental feeding to support populations.	2	4	3

| Translocate the population to a more suitable breeding area | Known to be beneficial in other seabird groups, but evidence for gulls is limited. Several failed attempts have been recorded, and to our knowledge no successful translocations of gulls have been carried out. | 3 | 4 | 4 |

Green = Likely to be beneficial. Red = Unlikely to be beneficial, may have negative impact. Orange = contradicting or uncertain evidence. Grey = Limited evidence.
R = relevance rating. S = strength rating. T = transparency rating. All ratings on a scale of 1 to 5, where 5 is the highest.

Details:

Artificially incubate or hand-rear chicks to support population
Relevance (R): 2 studies in the evidence base focus on gulls, 38 on other seabirds and 0 on other birds. **Strength (S):** The evidence base was comprised of 40 studies. Of these 9 were considered to have a good sample size, and 19 had a clear metric for effectiveness. **Transparency (T):** 26 studies included were published and peer-reviewed, 0 were from the grey literature, and 0 were anecdotal. Of the studies included, 17 had a published methodology, and 4 justified their rationale.

Make new colonies more attractive to encourage birds to colonise
Relevance (R): 0 studies in the evidence base focus on gulls, 38 on other seabirds and 6 on other birds. **Strength (S):** The evidence base was comprised of 44 studies. Of these 31 were considered to have a good sample size, and 18 had a clear metric for effectiveness. **Transparency (T):** 44 studies included were published and peer-reviewed, of which 1 were literature reviews or meta-analyses, 0 were from the grey literature, and 0 were anecdotal. Of the studies included, 30 had a published methodology, and 22 justified their rationale.

Provide supplementary food during the breeding season
Relevance (R): 4 studies in the evidence base focus on gulls, 12 on other seabirds and 0 on other birds. **Strength (S):** The evidence base was comprised of 16 studies. Of these 10 were considered to have a good sample size, and 14 had a clear metric for effectiveness. **Transparency (T):** 16 studies included were published and peer-reviewed, 0 were from the grey literature, and 0 were anecdotal. Of the studies included, 13 had a published methodology, and 4 justified their rationale.

Translocate the population to a more suitable breeding area

Relevance (R): 1 study in the evidence base focusses on gulls, 14 on other seabirds and 0 on other birds. **Strength (S):** The evidence base was comprised of 15 studies. Of these 13 were considered to have a good sample size, and 9 had a clear metric for effectiveness. **Transparency (T):** 14 studies included were published and peer-reviewed, of which 1 were literature reviews or meta-analyses, 0 were from the grey literature, and 0 were anecdotal. Of the studies included, 11 had a published methodology, and 9 justified their rationale.

6 Impact: Increased parasite load

Summary:

Treatment and prevention options are available for some parasites, but they are generally rarely trialled on seabirds, and the bulk of available knowledge is based on non-seabird species. Careful consideration and planning is needed before embarking on mass-treatment of seabird populations, to avoid unintended negative consequences.

Intervention	Evidence of Effectiveness	R	S	T
Inoculation or treatment against disease and parasites	Extensive literature exists for treatment of birds in general, but limited examples for seabirds and none for gulls. Many treatment and prevention options are available, but those that have been trialled have limited success, or even cause more harm than benefits, in wild seabird populations. The advisability of this action likely depends on the species and context in question. Endoparasite treatment in seabirds is particularly under-researched.	1	5	4

Green = Likely to be beneficial. Red = Unlikely to be beneficial, may have negative impact. Orange = contradicting or uncertain evidence. Grey = Limited evidence.
R = relevance rating. **S** = strength rating. **T** = transparency rating. All ratings on a scale of 1 to 5, where 5 is the highest.

Inoculation or treatment against disease and parasites

Relevance (R): 0 studies in the evidence base focus on gulls, 5 on other seabirds and 29 on other birds. **Strength (S):** The evidence base was comprised of 34 studies. Of these 25 were considered to have a good sample size, and 22 had a clear metric for effectiveness. **Transparency (T):** 34 studies included were published and peer-reviewed, of which 1 were literature reviews or meta-analyses, 0 were from the grey literature, and 0 were anecdotal. Of the studies included, 21 had a published methodology, and 26 justified their rationale.

7 Impact: Increased frequency/severity of storms (including wind, rain and wave action) causes nest destruction

Summary:

While there are several local actions that may prevent or mitigate local nest destruction, they have not been trialled widely and wide-spread evidence to support their use is currently lacking. If changes in extreme weather threatens the viability of a population, then several actions are available to encourage translocation of populations to safer areas.

Intervention	Evidence of Effectiveness	R	S	T
Alter habitat to encourage birds to leave an area	Few trials on seabirds and none on gulls. Several trials of this action have been successful and encouraged terns to shift breeding sites. However, this action is likely more viable for species with lower site fidelity and areas with other available breeding habitat nearby.	2	2	3
Artificially incubate or hand-rear chicks to support population	Known to be effective for some seabirds, though labour intensive and usually only appropriate for small populations. Gulls have been successfully hand-reared, but only in very small numbers. Likely to be difficult for many species, especially those that breed in steep, inaccessible habitats.	3	2	1
Install barriers to prevent flooding	While likely to prevent flooding there is currently no evidence available on this action's effectiveness in relation to seabird conservation	NA	NA	NA

Make new colonies more attractive to encourage birds to colonise	Several different methods have been trialled extensively across other seabirds, with variable success depending on method and species. No evidence currently available for gulls, the effectiveness of decoys, acoustic cues, smells and improved habitat is currently unknown.	2	4	3
Manually relocate nests	This has been reported by practitioners as an effective action to assist seabirds on low-lying beaches in the Baltic. However, to our knowledge there are no broad-scale studies or reviews of this action's effectiveness. The risk of disturbance is high, so is likely only an option as a last resort.	NA	NA	NA
Provide additional shelter or protection from extreme weather (flooding)	There are few trials on seabird species, most known examples are on terns, and most report little to no benefit for breeding populations. However, evidence is limited and more research is needed on this action's overall effectiveness. We found no published trials on gull species.	1	3	5
Provide artificial nesting sites	Tried extensively on many seabird species with significant benefit to many species. Artificial nesting sites have been successfully used to support kittiwake populations; for other gull species results have been mixed but several species have benefited from artificial nesting sites.	3	5	3
Repair/support nests to support breeding	Very limited evidence for effectiveness in seabirds, though known to be effective in other birds. No known examples in gull species.	2	2	3

Translocate the population to a more suitable breeding area	Known to be beneficial in other seabird groups, but evidence for gulls is limited. Several failed attempts have been recorded, and to our knowledge no successful translocations of gulls have been carried out.	3	4	4

Green = Likely to be beneficial. Red = Unlikely to be beneficial, may have negative impact. Orange = contradicting or uncertain evidence. Grey = Limited evidence. **R** = relevance rating. **S** = strength rating. **T** = transparency rating. All ratings on a scale of 1 to 5, where 5 is the highest.

Details:

Alter habitat to encourage birds to leave an area
Relevance (R): 0 studies in the evidence base focus on gulls, 2 on other seabirds and 0 on other birds. **Strength (S):** The evidence base was comprised of 2 studies. Of these 2 were considered to have a good sample size, and 0 had a clear metric for effectiveness. **Transparency (T):** 2 studies included were published and peer-reviewed, 0 were from the grey literature, and 0 were anecdotal. Of the studies included, 2 had a published methodology, and 1 justified their rationale.

Artificially incubate or hand-rear chicks to support population
Relevance (R): 2 studies in the evidence base focus on gulls, 38 on other seabirds and 0 on other birds. **Strength (S):** The evidence base was comprised of 40 studies. Of these 9 were considered to have a good sample size, and 19 had a clear metric for effectiveness. **Transparency (T):** 26 studies included were published and peer-reviewed, 0 were from the grey literature, and 0 were anecdotal. Of the studies included, 17 had a published methodology, and 4 justified their rationale.

Make new colonies more attractive to encourage birds to colonise
Relevance (R): 0 studies in the evidence base focus on gulls, 38 on other seabirds and 6 on other birds. **Strength (S):** The evidence base was comprised of 44 studies. Of these 31 were considered to have a good sample size, and 18 had a clear metric for effectiveness. **Transparency (T):** 44 studies included were published and peer-reviewed, of which 1 were literature reviews or meta-analyses, 0 were from the grey literature, and 0 were anecdotal. Of the studies included, 30 had a published methodology, and 22 justified their rationale.

Provide additional shelter or protection from extreme weather (flooding)
Relevance (R): 0 studies in the evidence base focus on gulls, 0 on other seabirds and 3 on other birds. **Strength (S):** The evidence base was comprised of 3 studies. Of these 1 was considered to have a good sample size, and 2 had a clear metric for

effectiveness. **Transparency (T):** 3 studies included were published and peer-reviewed, 0 were from the grey literature, and 0 were anecdotal. Of the studies included, 3 had a published methodology, and 3 justified their rationale.

Provide artificial nesting sites
Relevance (R): 1 study in the evidence base focusses on gulls, 51 on other seabirds and 1 on other birds. **Strength (S):** The evidence base was comprised of 54 studies. Of these 50 were considered to have a good sample size, and 33 had a clear metric for effectiveness. **Transparency (T):** 53 studies included were published and peer-reviewed, of which 2 were literature reviews or meta-analyses, 0 were from the grey literature, and 0 were anecdotal. Of the studies included, 33 had a published methodology, and 27 justified their rationale.

Repair/support nests to support breeding
Relevance (R): 0 studies in the evidence base focus on gulls, 2 on other seabirds and 1 on other birds. **Strength (S):** The evidence base was comprised of 3 studies. Of these 1 was considered to have a good sample size, and 1 had a clear metric for effectiveness. Transparency (T): 3 studies included were published and peer-reviewed, 0 were from the grey literature, and 0 were anecdotal. Of the studies included, 1 had a published methodology, and 3 justified their rationale.

Translocate the population to a more suitable breeding area
Relevance (R): 1 study in the evidence base focusses on gulls, 14 on other seabirds and 0 on other birds. **Strength (S):** The evidence base was comprised of 15 studies. Of these 13 were considered to have a good sample size, and 9 had a clear metric for effectiveness. **Transparency (T):** 14 studies included were published and peer-reviewed, of which 1 were literature reviews or meta-analyses, 0 were from the grey literature, and 0 were anecdotal. Of the studies included, 11 had a published methodology, and 9 justified their rationale.

8 Impact: Increased frequency/severity of storms (including wind, rain and wave action) increases foraging difficulty and/or mortality

Summary:
Several local actions may be possible to limit mortality or increase recovery on a small scale, but for larger populations effective local action is difficult. Supporting the population in more general ways (increasing adult survival, limiting chick mortality) may be the most effective method.

Intervention	Evidence of Effectiveness	R	S	T
Provide supplementary food during the breeding season	Trialled on many seabird species. Known to be beneficial for several gull species, but success varies. Many gulls will scavenge any available food source, so it is feasible to provide supplementary food. However, as many gull populations are already reliant on discards, there are ethical concerns regarding wide-spread use of supplemental feeding to support populations.	3	4	3
Provide supplementary food during the non-breeding season	This is a hypothetical action. We found no published studies assessing this action's effectiveness for seabirds. While possible, especially for species that remain near land, the same concerns apply as during the breeding season. Many gull populations are already reliant on discards, and there are ethical concerns regarding wide-spread use of supplemental feeding to support populations.	NA	NA	NA
Rehabilitate sick or injured birds	For groups of long-lived, large birds, rehabilitation is known to be an effective way to support populations. However, examples in seabirds are scarce and the overall effectiveness for most species is unknown. However, there are numerous successful reports of rehabilitation and release in various gull species from rescue centres. Likely a feasible action, at least at small numbers of individuals.	1	2	4

Green = Likely to be beneficial. Red = Unlikely to be beneficial, may have negative impact.
Orange = contradicting or uncertain evidence. Grey = Limited evidence.
R = relevance rating. S = strength rating. T = transparency rating. All ratings on a scale of 1 to 5, where 5 is the highest.

Details:

Provide supplementary food during the breeding season

Relevance (R): 4 studies in the evidence base focus on gulls, 12 on other seabirds and 0 on other birds. **Strength (S):** The evidence base was comprised of 16 studies. Of these 10 were considered to have a good sample size, and 14 had a clear metric for effectiveness. **Transparency (T):** 16 studies included were published and peer-reviewed, 0 were from the grey literature, and 0 were anecdotal. Of the studies included, 13 had a published methodology, and 4 justified their rationale.

Rehabilitate sick or injured birds

Relevance (R): 0 studies in the evidence base focus on gulls, 3 on other seabirds and 4 on other birds. **Strength (S):** The evidence base was comprised of 7 studies. Of these 4 were considered to have a good sample size, and 1 had a clear metric for effectiveness. **Transparency (T):** 7 studies included were published and peer-reviewed, 0 were from the grey literature, and 0 were anecdotal. Of the studies included, 5 had a published methodology, and 5 justified their rationale.

© Silviu Petrovan

Loons/Divers and Grebes
(Gaviidae and Podicipedidae)

An assessment of climate change vulnerability and potential conservation actions for loons/divers and grebes in the North-East Atlantic

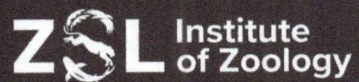

UNIVERSITY OF CAMBRIDGE

ZSL Institute of Zoology

https://doi.org/10.11647/OBP.0343.05

1 Arctic Loon (*Gavia arctica*)

1.1 Evidence for exposure

1.1.1 Potential changes in breeding habitat suitability (by 2100):

🟥 Current breeding area that is likely to become less suitable (96% of current range).

🟨 Current breeding area that is likely to remain suitable (3%).

🟩 Current breeding area that is likely to become more suitable (1%).

1.1.2 Current impacts attributed to climate change:

We did not identify any current impacts of climate change for this species.

1.1.3 Predicted changes in key prey species:

① Key prey species are likely to decline in abundance along the Swedish Baltic coast.

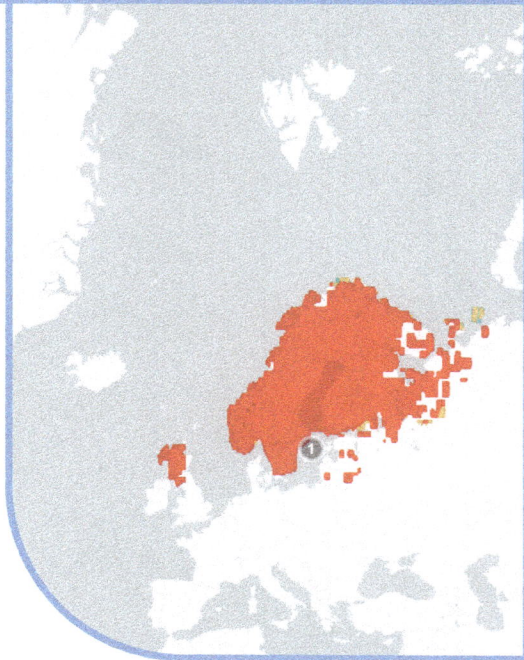

1.2 Sensitivity

• Loons have a strong preference for breeding near nutrient-poor, very clear fresh water. There is strong evidence that the productivity of lakes is increasing across Fenno-Scandinavia, which has resulted in decreased water transparency, and therefore decreased foraging success and condition of loon chicks. "Browning" of fresh water lakes has also led to reduced biomass of key fish species, but also a shift towards smaller fish, which could potentially result in positive or negative effects for loons. Climate change is likely to accelerate this process and could have significant impacts on loons in the future.

• Species is sensitive to many threats, including disturbance by forestry work and tourism, ship traffic, bycatch and wind farms. Nest abandonment is common when disturbed. Conservation intervention may therefore be difficult.

• Sensitive to flooding due to storms and increased wave action as loons tend to nest close to water level in exposed areas. Any water level rise or increase in wave action is likely to have significant impact on breeding populations. Nests are also vulnerable to flooding due to heavy precipitation, and as significant rain events are expected to increase in many areas due to climate change, flooding events could become more frequent.

1.3 Adaptive capacity

• No relevant information could be identified.

Note: loons/divers and grebes rely heavily on freshwater prey and habitats. While these were considered for this assessment, we focused on marine habitats and prey. Estimates of breeding habitat and prey loss should therefore be treated as incomplete.

© Seppo Häkkinen

2 Common Loon *(Gavia immer)*

1.1 Evidence for exposure

1.1.1 Potential changes in breeding habitat suitability (by 2100):

■ Current breeding area that is likely to become less suitable (96% of current range).

■ Current breeding area that is likely to remain suitable (2%).

■ Current breeding area that is likely to become more suitable (2%).

1.1.2 Current impacts attributed to climate change:

We did not identify any current impacts of climate change for this species.

1.1.3 Predicted changes in key prey species:

No key prey assessment was carried out for this species.

1.1.4 Climate change impacts outside of Europe

Several impacts of climate change have been noted in North American populations, including decreased brood size, changes in migration patterns, increased energetic stress due to higher temperatures, and an increase in exposure to mercury.

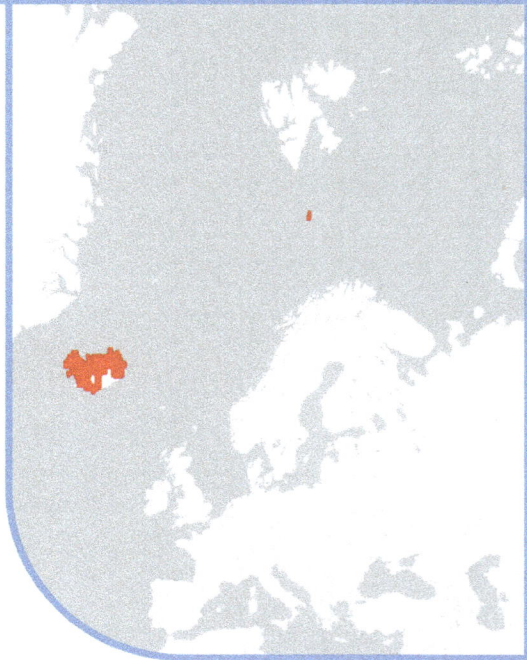

1.2 Sensitivity

• Loon nests are vulnerable to flooding given their placement at the shoreline's edge. As significant rain events are expected to increase in many areas due to climate change, increased flooding could have significant impacts.

• This species has a long generation length (>10 years), which may slow recovery from severe impacts and increases population extinction risk.

1.3 Adaptive capacity

• Feeds on a wide variety of marine and freshwater fish, and appears to prey-switch readily. Loss of one or a few species due to climate change is unlikely to significantly impact most populations.

• Common loons frequently return to the same breeding sites, and show strong breeding site fidelity. In addition, dispersal is low so new areas are not commonly colonised. However, following repeated breeding failures several populations have abandoned previous sites and colonised new areas. This suggests some populations may shift in response to changing climate.

• Common loons have high fidelity to wintering sites, and they are unlikely to respond quickly to change in conditions in these wintering areas. Any major change to these areas may have significant impacts on wintering populations.

Note: loons/divers and grebes rely heavily on freshwater prey and habitats. While these were considered for this assessment, we focused on marine habitats and prey. Estimates of breeding habitat and prey loss should therefore be treated as incomplete.

Common loon © Silviu Petrovan

3 Red-throated Loon *(Gavia stellata)*

1.1 Evidence for exposure

1.1.1 Potential changes in breeding habitat suitability (by 2100):

■ Current breeding area that is likely to become less suitable (93% of current range).

■ Current breeding area that is likely to remain suitable (4%).

■ Current breeding area that is likely to become more suitable (2%).

1.1.2 Current impacts attributed to climate change:

We did not identify any current impacts of climate change for this species.

1.1.3 Predicted changes in key prey species:

① Key prey species are likely to decline in abundance along the south-western Swedish coast and the Baltic Sea.

1.1.4 Climate change impacts outside of Europe

• Several impacts of climate change have been noted in North American populations, including decreased brood size, changes in migration patterns, increased energetic stress due to higher temperatures, and an increase in exposure to mercury.

1.2 Sensitivity

• Loons have a strong preference for breeding near nutrient-poor, very clear

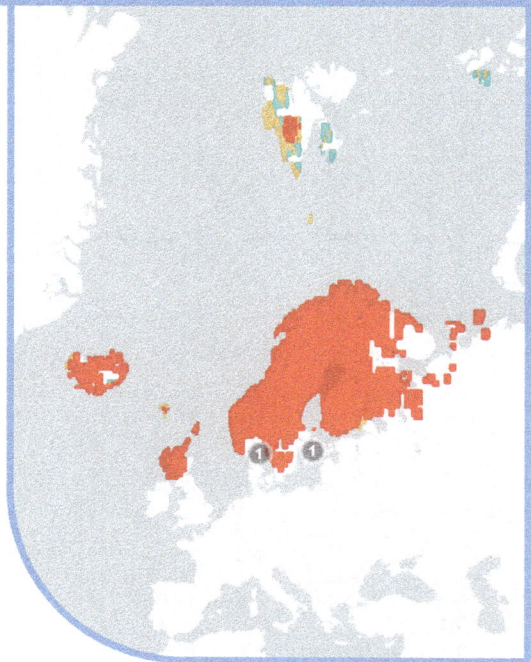

fresh water. Productivity of lakes is increasing across Fenno-Scandinavia, which has resulted in decreased water transparency, and therefore decreased foraging success and condition of loon chicks. "Browning" of fresh water lakes has also led to reduced biomass and smaller body size of key fish species, which have both positive and negative consequences for loons.

• Some populations of red-throated loons (particularly in Iceland) rely on small, shallow ponds for feeding and for nesting sites. In hot, dry summers these pools can dry up and result in difficulty in take-off, foraging and thermoregulation. Further hotter, drier summers caused by climate change could have significant impacts on these populations.

• Climate change at high latitudes could greatly reduce the area of suitable breeding habitat for red-throated loons that breed in these areas. Warming can lead to changes in lake ice-coverage and drainage, which in turn can lead to the temporary or permanent loss of key tundra habitat for breeding loons.

• Red-throated loons are sensitive to disturbance, human activity can result in many nests being abandoned. Conservation intervention during the breeding season may therefore be difficult to carry out without disturbing nesting birds.

• Red-throated loons are also sensitive to predation, low overall breeding success rate is often due to heavy predation. Change in predator range or abundance due to climate change is likely to have significant impacts on loons.

• Nests are sensitive to flooding following high rainfall or heavy storms. Any increase in extreme events could significantly impact breeding populations.

1.3 Adaptive capacity

• Red-throated loons frequently return to the same breeding sites, and show some breeding site fidelity. However, when conditions change several populations have abandoned previous sites and colonised new areas. This suggests some populations may shift in response to changing climate.

• Red-throated loons have high fidelity to wintering sites, and they are unlikely to respond quickly to change in conditions in these wintering areas. Any major change to these areas may have significant impacts on wintering populations.

• Migration of red-throated loons often varies year to year in response to wind and weather conditions, though flexibility is likely to be limited due to complex interaction of carry-over effects and role of staged migration. Loons may be able to change migration strategy to some extent in response to climate change.

Note: loons/divers and grebes rely heavily on freshwater prey and habitats. While these were considered for this assessment, we focused on marine habitats and prey. Estimates of breeding habitat and prey loss should therefore be treated as incomplete.

4 Horned Grebe *(Podiceps auritus)*

1.1 Evidence for exposure

1.1.1 Potential changes in breeding habitat suitability (by 2100):

■ Current breeding area that is likely to become less suitable (92% of current range).

■ Current breeding area that is likely to remain suitable (7%).

■ Current breeding area that is likely to become more suitable (1%).

1.1.2 Current impacts attributed to climate change:

We did not identify any current impacts of climate change for this species.

1.1.3 Predicted changes in key prey species:

No key prey assessment was carried out for this species.

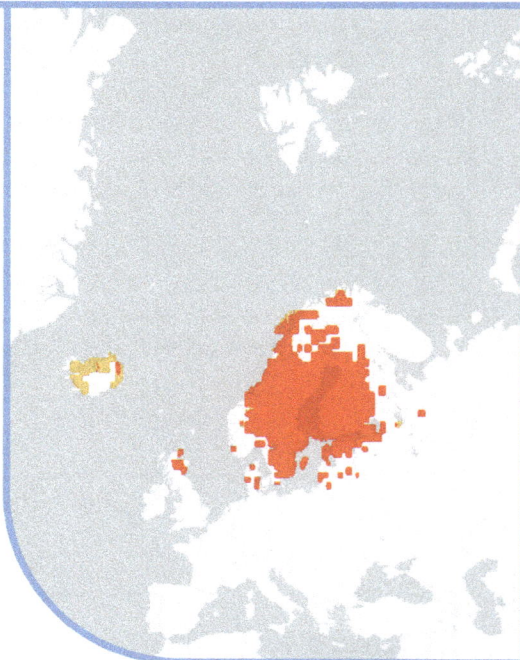

1.2 Sensitivity

• The species has a large range, and a large population (though the population in Europe is relatively small), but is declining rapidly across much of its range. The cause is unknown, previous research has found little to no evidence climate change is contributing to decline. However, additional pressure from climate change is likely to exacerbate existing declines.

• Species is sensitive to disturbance, sites or whole areas can be abandoned if there is heavy human activity. Conservation intervention or monitoring is therefore difficult.

• Horned grebes build either floating nests or nests on low-lying rocks and beaches. Nests are therefore vulnerable to flooding caused by high wind and wave action, it can be a significant source of mortality. Any changes in lake or coastal sea level are likely to have impacts on breeding populations.

1.3 Adaptive capacity

• Horned grebes frequently return to the same breeding sites, and show some breeding site fidelity. However, when conditions change several populations have abandoned previous sites. There appears to be significant interchange between populations, and some populations may shift in response to changing climate.

• Species has a varied diet of freshwater and marine species. Change in availability of one or a few species is unlikely to have significant impacts on most populations.

Note: loons/divers and grebes rely heavily on freshwater prey and habitats. While these were considered for this assessment, we focused on marine habitats and prey. Estimates of breeding habitat and prey loss should therefore be treated as incomplete.

5 Red-necked Grebe
(Podiceps grisegena)

1.1 Evidence for exposure

1.1.1 Potential changes in breeding habitat suitability (by 2100):

■ Current breeding area that is likely to become less suitable (98% of current range).

■ Current breeding area that is likely to remain suitable (1%).

■ Current breeding area that is likely to become more suitable (0%).

1.1.2 Current impacts attributed to climate change:

We did not identify any current impacts of climate change for this species.

1.1.3 Predicted changes in key prey species:

No key prey assessment was carried out for this species.

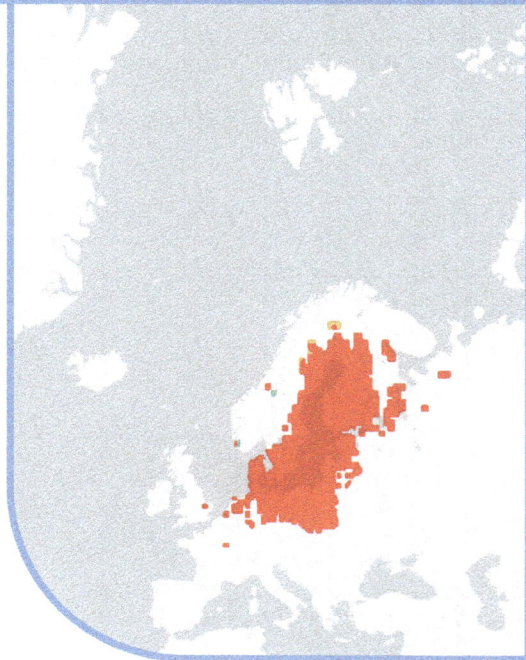

1.2 Sensitivity

• Sensitive to disturbance; nests are frequently abandoned in presence of human activity. May make conservation action or monitoring more difficult.
• Sensitive to flooding and wave action as nest sites are often near the water-line and exposed. In some years this is a major cause of nest and egg loss, and any increase in sea level or frequency of storms may have significant impacts on nesting sites.

1.3 Adaptive capacity

• Red-necked grebes readily change breeding site, often changing their site each year. In some cases they will even change nest site during the breeding season. They could potentially rapidly change breeding areas in response to climate change.

• Has a varied diet of both fresh and saltwater species. The loss of one or a few species is unlikely to significantly impact most populations.

Note: loons/divers and grebes rely heavily on freshwater prey and habitats. While these were considered for this assessment, we focused on marine habitats and prey. Estimates of breeding habitat and prey loss should therefore be treated as incomplete.

© Seppo Häkkinen

Potential actions in response to climate change: Loons/ Divers and Grebes (Gaviidae and Podicipedidae)

In this section we list and assess possible local conservation actions that could be carried out in response to identified climate change impacts on auks . This section is not grouped by species, but by identified impacts. If an impact or action is specific to one or a few species, this information is included in the action summary or in the footnotes.

Additional note: loons/divers and grebes rely heavily on fresh water prey and habitats. While these were considered for this assessment, we focussed heavily on marine habitats and prey.

We did not identify any current impacts of climate change for this group.

© Seppo Häkkinen

© Seppo Häkkinen

Petrels and Shearwaters

(Hydrobatidae and Procellariidae)

An assessment of climate change vulnerability and potential conservation actions for petrels and shearwaters in the North-East Atlantic

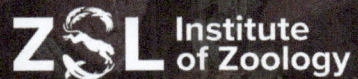

UNIVERSITY OF CAMBRIDGE

ZSL Institute of Zoology

https://doi.org/10.11647/OBP.0343.06

1 Cory's Shearwater

(Calonectris borealis)

1.1 Evidence for exposure

1.1.1 Potential changes in breeding habitat suitability (by 2100):

🟧 Current breeding area that is likely to become less suitable (100% of current range).

🟨 Current breeding area that is likely to remain suitable (0%).

🟩 Current breeding area that is likely to become more suitable (0%).

1.1.2 Current impacts attributed to climate change:

① **Neutral Impact:** New colonies have been established outside of the species' normal range. The cause is uncertain, but likely related to prey range shifts and warming conditions.

1.1.3 Predicted changes in key prey species:

No key prey species are predicted to decline for this species.

1.2 Sensitivity

• Cory's shearwaters can suffer mass-mortalities, especially following extreme storms or hurricanes in tropical parts of their range. Changes in how and where hurricanes occur could have significant impacts on shearwaters.

• Extreme positive and negative NAO indices drastically impact adult foraging patterns, adult condition, and chick condition, very likely because of changes in prey availability. Extreme fluctuations are likely to become more common in the future, and therefore may heavily impact shearwater breeding populations.

• Cory's shearwaters rely on wind conditions to soar and use as little energy as possible. Changes in wind strength, direction and patterns could heavily impact energy use and migration paths.

• Cory's shearwaters have high breeding synchrony; the majority of birds in a population will lay their eggs in a short period of time. If temporal shifts in key prey availability occur it could have a significant impact on breeding populations.

• This species has a long generation length (>10 years), which may slow recovery from severe impacts and increases population extinction risk.

1.3 Adaptive capacity

• Analysis of laying dates has shown that, regardless of sea temperatures and weather conditions, there is little variance in Cory's shearwater phenology. It is unlikely they will adapt their laying date to changing conditions.

• Cory's shearwater shows little plasticity in choosing nesting sites, and can even choose maladaptively in the presence of predators. Although it recently has established in northern Spain, this is believed to be a rare event. High site fidelity means it is unlikely to change sites as a response to climate change.

• Cory's shearwater change their foraging strategy and prey species based on prey availability. This, combined with a relatively flexible diet, is likely to make them more resilient to climate change.

2 Northern Fulmar (*Fulmarus glacialis*)

1.1 Evidence for exposure

1.1.1 Potential changes in breeding habitat suitability (by 2100):

🟧 Current breeding area that is likely to become less suitable (80% of current range).

🟨 Current breeding area that is likely to remain suitable (20%).

🟩 Current breeding area that is likely to become more suitable (0%).

1.1.2 Current impacts attributed to climate change:

① **Neutral Impact:** Warmer winters have resulted in lower adult survival and lower reproductive success in the following year. Mechanism uncertain, but could be related to marine productivity or to frequency and severity of storms.

② **Negative Impact:** Higher sea temperatures typically correlate with lower breeding success. Mechanism unknown, but likely mediated through prey availability. Continued warming may cause long-term declines in populations.

1.1.3 Predicted changes in key prey species:

③ Key prey species are likely to decline in abundance in the Irish Sea, the English Channel, the southern coast of Norway and along the Brittany coast.

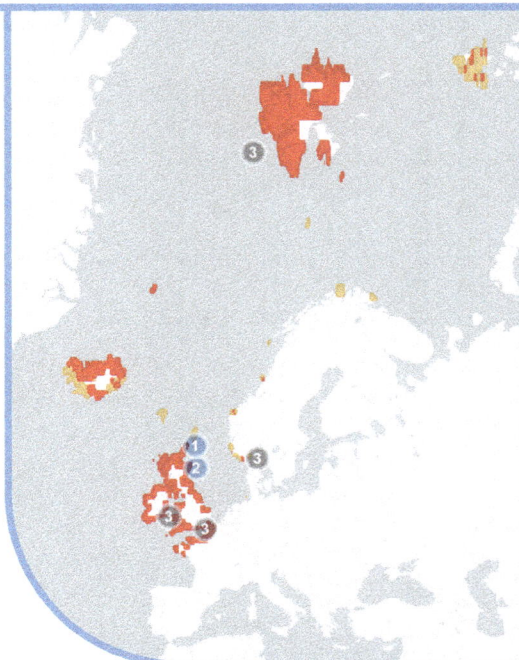

1.2 Sensitivity

• Fulmars are prone to wrecks across their range. Previous wrecks have been attributed to a combination of sustained high winds, unusually warm waters, and reduced food availability, all of which are more likely to occur more frequently due to climate change.

• Fulmars rely on wind conditions to soar and use as little energy as possible. Changes in wind strength, direction and patterns could heavily impact energy use and migration paths.

• Storms are also known to have some negative effects on fulmar breeding success as they lead to more difficult foraging conditions and lower body condition. Changes in storm patterns may affect fulmar breeding success, as well as contribute to wrecks.

• This species has a long generation length (>10 years), which may slow recovery from severe impacts and increases population extinction risk.

• Fulmars have high breeding synchrony; the majority of birds in a population will lay their eggs in a short period of time. If temporal shifts in key prey availability occur it could have significant impacts on breeding population.

1.3 Adaptive capacity

• Atlantic population has expanded dramatically in range and number during last 200 years, likely driven by increase in fishery discards. Species readily colonises new areas if they are suitable, which is likely to help fulmars adapt to climate change.

• Juveniles very frequently disperse to other colonies, but once breeding the species has very high site fidelity. Although new colonies are formed readily, existing populations are very unlikely to relocate.

• Very variable diet, even within comparatively small area. Different populations prey on very different species, presumably in response to availability. This is likely to buffer impacts of climate change and changes in marine ecosystems.

• Fulmars frequently skip breeding in poor conditions; this may be adaptive in response to climate change as it allows adults to maximise condition for good breeding years.

• Fulmars have shifted their migration timing and laying date in some populations; this may be related to conditions in breeding and non-breeding areas but the underlying reasons are currently unknown.

3 Band-rumped Storm-petrel

(Hydrobates castro)

1.1 Evidence for exposure

1.1.1 Potential changes in breeding habitat suitability (by 2100):

■ Current breeding area that is likely to become less suitable (100% of current range).

■ Current breeding area that is likely to remain suitable (0%).

■ Current breeding area that is likely to become more suitable (0%).

1.1.2 Current impacts attributed to climate change:

We did not identify any current impacts of climate change for this species.

1.1.3 Predicted changes in key prey species:

No key prey species are predicted to decline for this species.

1.2 Sensitivity

• Species is declining in many areas; range was likely much larger historically but has been greatly reduced by introduced predators. Any additional pressure from climate change is likely to accelerate these declines

• Band-rumped storm-petrels rely on wind conditions to soar and use as little energy as possible. Changes in wind strength, direction and patterns could heavily impact energy use and migration paths.

• Band-rumped storm-petrels have high breeding synchrony; the majority of birds in a population will lay their eggs in a short period of time. If temporal shifts in key prey availability occur it could have significant impacts on breeding populations

• This species has a long generation length (>10 years), which may slow recovery from severe impacts and increases population extinction risk.

1.3 Adaptive capacity

• Species has very high fidelity to breeding areas, and even to individual burrows. Unlikely to disperse and colonise new areas readily in response to climate change.

• There is very limited mixing between populations in storm-petrels. This can be adaptive (as it increases likelihood of local adaptation) or non-adaptive (as immigration to support populations is unlikely).

Northern fulmar chick © Seppo Häkkinen

4 Leach's Storm-petrel

(Hydrobates leucorhous)

1.1 Evidence for exposure

1.1.1 Potential changes in breeding habitat suitability (by 2100):

■ Current breeding area that is likely to become less suitable (76% of current range).

■ Current breeding area that is likely to remain suitable (24%).

■ Current breeding area that is likely to become more suitable (0%).

1.1.2 Current impacts attributed to climate change:

We did not identify any current impacts of climate change for this species.

1.1.3 Predicted changes in key prey species:

No key prey species are predicted to decline for this species.

1.1.4 Climate change impacts outside of Europe

• Leach's storm-petrels in North America have changed their prey species and foraging strategy in response to shifts in the marine ecosystem partially driven by climate change.

• Heatwaves in North America have impacted storm-petrel colonies and resulted in changes in diet, loss of condition and wrecks.

• Leach's storm-petrel reproductive success has been linked to global temperature. Warmer temperatures result in higher reproductive success, up until a certain threshold after which it decreases. The mechanism is unknown.

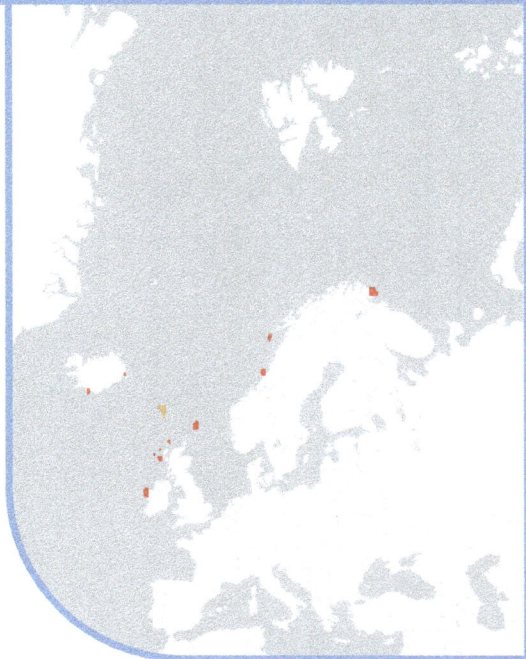

1.2 Sensitivity

• Species has a large population and large range but is rapidly declining across its range. The exact cause is not certain, but there are likely a number of factors, including bycatch, avian and mammal predation, pollution and disturbance. Any additional pressure from climate change is likely to accelerate these declines.

• Known to wreck in great numbers in various parts of its range, especially when strong prolonged off-shore winds occur, which can blow petrels far from their usual migration route. Changes in wind and storm patterns could potentially have significant impacts.

• High temperatures in other parts of the species' range, especially heat-waves, are associated with lower prey availability, loss of condition, lower reproductive output and sometimes with die-offs. An increase in temperature or heatwaves is likely to have a significant impact on breeding colonies.

• Leach's storm-petrels rely on wind conditions to soar and use as little energy as possible. Changes in wind strength, direction and patterns could heavily impact energy use and migration paths.

• This species has a long generation length (>10 years), which may slow recovery from severe impacts and increases population extinction risk.

1.3 Adaptive capacity

• Leach's storm-petrel has a flexible migration strategy, and changes its migration route, stop-overs and wintering areas based on conditions. Local changes to migration routes are unlikely to have a major impact.

• Leach's storm-petrel is known to change prey species and foraging areas in response to changes in conditions. They have a very large foraging range and local changes in climate are therefore less likely to have a large impact.

• Analysis of laying dates has shown that, regardless of SST and weather conditions, there is little variance in shearwater phenology. It is unlikely they will adapt their laying date to changing conditions.

• Species has very high fidelity to breeding areas. Unlikely to disperse and colonise new areas readily.

• There is very limited mixing between populations in storm-petrels. This can be adaptive (as it increases likelihood of local adaptation) or non-adaptive (as immigration to support populations is unlikely).

5 European Storm-petrel

(Hydrobates pelagicus)

1.1 Evidence for exposure

1.1.1 Potential changes in breeding habitat suitability (by 2100):

■ Current breeding area that is likely to become less suitable (72% of current range).

■ Current breeding area that is likely to remain suitable (28%).

■ Current breeding area that is likely to become more suitable (0%).

1.1.2 Current impacts attributed to climate change:

1 **Negative Impact:** High winds and storms in the non-breeding season cause increased mortality, lower body condition, and reduced breeding success. While individual extreme climate events are difficult to attribute to climate change, it is likely that climate change is driving an increase in their frequency and/or intensity.

2 **Neutral Impact:** A shift towards warmer, drier and calmer conditions has resulted in lower storm-petrel abundance. The mechanism is unknown, but likely related to changes in marine ecosystem and key prey availability.

1.1.3 Predicted changes in key prey species:

No key prey species are predicted to decline for this species.

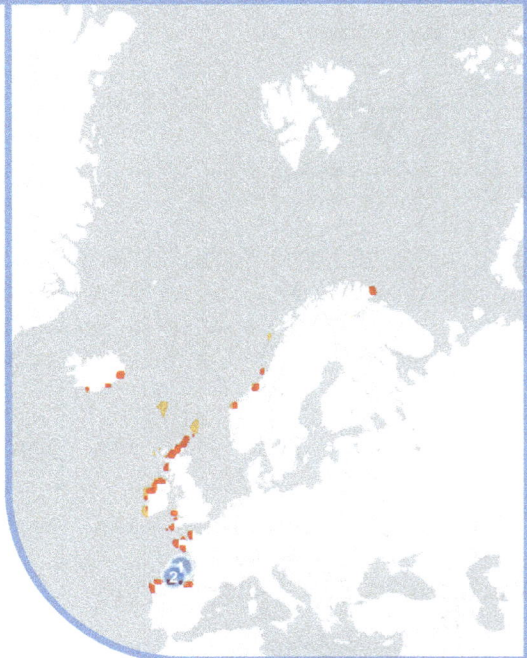

1.2 Sensitivity

• Studies in other parts of this species' range have found it is sensitive to high rainfall and low temperatures, which causes high chick mortality. If climate change causes periods of low temperatures and higher rainfall during the breeding season, this may heavily impact juvenile survival.

• Mortality of breeding storm-petrels is significantly increased during summers with a high incidence of heavy storms, as they create difficult foraging conditions and lower body condition. Projected increase in the frequency of storms may therefore heavily impact storm-petrel populations.

• European storm-petrels rely on wind patterns to fly and navigate; at high wind speeds storm-petrels are vulnerable to being storm-driven. Higher incidence of extreme wind events is likely to have significant impacts on foraging ability and marine distribution.

• This species has a long generation length (>10 years), which may slow recovery from severe impacts and increase population extinction risk.

1.3 Adaptive capacity

• Analysis of laying dates has shown that, regardless of sea temperatures and weather conditions, there is little variance in storm-petrel phenology. It is unlikely they will adapt their laying date to changing conditions, which could make them vulnerable to climate change. However, additional evidence has found that changes in marine temperature do not affect breeding success for storm-petrels in the Mediterranean.

• In response to poor breeding conditions, European storm-petrels can skip breeding events. This could buffer the negative effects of climate change, as storm-petrels could maintain body condition even in poor years by skipping breeding.

• Species has very high fidelity to breeding areas. Unlikely to disperse and colonise new areas readily. However, there is anecdotal evidence that European storm-petrels will recolonise old breeding sites if key threats are removed (e.g. predators).

• European storm-petrels have flexible foraging behaviour and prey on a wide variety of species. They are likely to take advantage of alternative prey sources if climate change impacts one or a few prey species.

• There is very limited mixing between populations in storm-petrels. This can be adaptive (as it increases likelihood of local adaptation) or non-adaptive (as immigration to support populations is unlikely).

6 Manx Shearwater *(Puffinus puffinus)*

1.1 Evidence for exposure

1.1.1 Potential changes in breeding habitat suitability (by 2100):

🟥 Current breeding area that is likely to become less suitable (66% of current range)

🟨 Current breeding area that is likely to remain suitable (34%)

🟩 Current breeding area that is likely to become more suitable (0%)

1.1.2 Current impacts attributed to climate change:

1 **Negative Impact:** Reduced prey availability during the breeding season has led to longer foraging trips and lower condition in adults and chicks.

1.1.3 Predicted changes in key prey species:

2 Key prey species are likely to decline in abundance in the Irish Sea.

1.1.4 Climate change impacts outside of Europe

• Manx shearwaters are known to be sensitive to climate change in the tropics, particularly to wrecks caused by storms, which are becoming more common due to changes in the El Niño cycle.

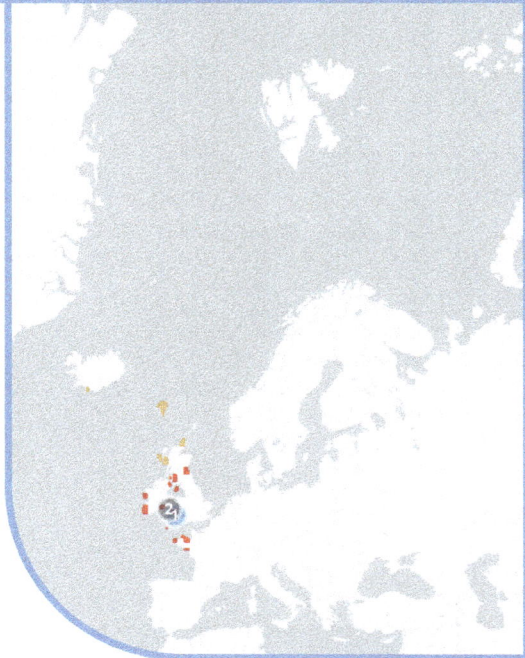

1.2 Sensitivity

• Manx shearwater burrows are prone to flooding; heavy rainfall during the incubation period can result in many nests being lost. Some populations may even be constrained by the number of nest sites that are prone to flooding. An increase in extreme precipitation events due to climate change is likely to have significant impacts on shearwater populations.

• Manx shearwaters, and shearwaters in general, rely on wind conditions to soar and use as little energy as possible. Changes in wind strength, direction and patterns could heavily impact energy use and migration paths.

• This species has a long generation length (>10 years), which may slow recovery from severe impacts and increase population extinction risk.

1.3 Adaptive capacity

• Juveniles occasionally disperse to other colonies, but once breeding the species has very high fidelity. There is limited mixing between groups, leading to population divergence, and they are unlikely to disperse to or found new colonies.

• Despite the species' generally high site fidelity, Manx shearwaters have colonised parts of North America in the last century, indicating the species can successfully establish new colonies. Climate change is potentially a contributing factor to this shift in distribution. However, this is considered a relatively rare event, and it seems unlikely the species can rapidly shift their distribution in response to climate change.

Northern fulmar © Silviu Petrovan

Potential actions in response to climate change: Petrels and Shearwaters (Hydrobatidae and Procellariidae)

In this section we list and assess possible local conservation actions that could be carried out in response to identified climate change impacts. This section is not grouped by species, but by identified impacts. If an impact or action is specific to one or a few species, this information is included in the action summary or in the footnotes. Effectiveness, relevance, strength and transparency scores are based on the available evidence we collated (see Appendix 2), and therefore all statements regarding limited or a lack of evidence relate to the collated evidence base, and does not infer that no such studies exist.

1 Impact: Increased frequency/severity of storms (including wind, rain and wave action) increases foraging difficulty and/or mortality

Summary:
Invasive mammals are a major threat to many seabird populations, and as such there is a well-established literature on mammal exclusion, management and eradication detailing effective methods and case studies. However, there are more limited options when the mammalian predator in question is itself a conservation target, or is not easily managed. Nevertheless, for many situations there are several, well-researched, actions available that can benefit seabird populations effectively.

Intervention	Evidence of Effectiveness	R	S	T
Provide supplementary food during the breeding season	Trialled on many seabird species. Very limited trials on petrels and shearwaters, and with limited success. It may be possible to feed a small number of chicks for a limited amount of time, but feeding adults supplementary food is likely unfeasible or, at the least, extremely challenging.	3	4	3
Provide supplementary food during the non-breeding season	This is a hypothetical action. We found no published studies assessing this action's effectiveness for seabirds.	NA	NA	NA
Rehabilitate sick or injured birds	For groups of long-lived, large birds, rehabilitation is known to be an effective way to support populations. However, examples in seabirds are scarce and the overall effectiveness for most species is unknown. Several species of petrels and shearwaters have been rescued and rehabilitated, but success rates are generally very low. Many species are prone to respiratory problems unless release is quick.	1	2	4

Green = Likely to be beneficial. Red = Unlikely to be beneficial, may have negative impact.
Orange = contradicting or uncertain evidence. Grey = Limited evidence.
R = relevance rating. S = strength rating. T = transparency rating. All ratings on a scale of 1 to 5, where 5 is the highest.

Details:

Provide supplementary food during the breeding season
Relevance (R): 2 studies in the evidence base focus on petrels and shearwaters, 14 on other seabirds and 0 on other birds. **Strength (S):** The evidence base was comprised of 16 studies. Of these 10 were considered to have a good sample size, and 14 had a clear metric for effectiveness. **Transparency (T):** 16 studies included were published and peer-reviewed, 0 were from the grey literature, and 0 were anecdotal. Of the studies included, 13 had a published methodology, and 4 justified their rationale.

Rehabilitate sick or injured birds

Relevance (R): 0 studies in the evidence base focus on petrels and shearwaters, 3 on other seabirds and 4 on other birds. **Strength (S):** The evidence base was comprised of 7 studies. Of these 4 were considered to have a good sample size, and 1 had a clear metric for effectiveness. **Transparency (T):** 7 studies included were published and peer-reviewed, 0 were from the grey literature, and 0 were anecdotal. Of the studies included, 5 had a published methodology, and 5 justified their rationale.

2 Impact: Reduced prey availability during breeding season

Summary:

Several local actions may assist breeding populations on a small scale, but direct intervention on a large scale is likely to be extremely difficult. General conservation actions to protect fish stocks and local marine areas may be the most effective method. If a population is likely to suffer major losses, even with conservation help, then translocations could be considered.

Intervention	Evidence of Effectiveness	R	S	T
Artificially incubate or hand-rear chicks to support population	Known to be effective for some seabirds, though labour intensive and usually only appropriate for small populations. Several petrel and shearwater species have been hand-reared successfully, but typically in small numbers. Longer-term ex-situ populations are likely to be unfeasible.	3	2	3
Make new colonies more attractive to encourage birds to colonise	Trialled extensively across other seabirds, with variable success. However, in petrels and shearwaters the use of vocalisations, smells and suitable burrows have been generally successful in encouraging establishment, especially when combined with other conservation actions. The effectiveness of each individual action is mixed, and varies depending on the species and population in question.	2	4	3

		R	S	T
Provide supplementary food during the breeding season	Trialled on many seabird species. Very limited trials on petrels and shearwaters, and with limited success. It may be possible to feed a small number of chicks for a limited amount of time, but feeding adults supplementary food is likely unfeasible or, at the least, extremely challenging.	3	4	3
Translocate the population to a more suitable breeding area	Known to be beneficial in several seabird groups, and multiple translocations of petrels and shearwaters have been carried out successfully. There is a substantial body of work on maximising translocation success, and encouraging establishment. Note however translocations have been in the context of island restoration and predator removal, and (like many other actions) have not been trialled as a response to climate change.	4	4	4

Green = Likely to be beneficial. Red = Unlikely to be beneficial, may have negative impact. Orange = contradicting or uncertain evidence. Grey = Limited evidence.
R = relevance rating. S = strength rating. T = transparency rating. All ratings on a scale of 1 to 5, where 5 is the highest.

Details:

Artificially incubate or hand-rear chicks to support population
Relevance (R): 4 studies in the evidence base focus on petrels and shearwaters, 36 on other seabirds and 0 on other birds. **Strength (S):** The evidence base was comprised of 40 studies. Of these 9 were considered to have a good sample size, and 19 had a clear metric for effectiveness. **Transparency (T):** 26 studies included were published and peer-reviewed, 0 were from the grey literature, and 0 were anecdotal. Of the studies included, 17 had a published methodology, and 4 justified their rationale.

Make new colonies more attractive to encourage birds to colonise
Relevance (R): 9 studies in the evidence base focus on petrels and shearwaters, 29 on other seabirds and 6 on other birds. **Strength (S):** The evidence base was comprised of 44 studies. Of these 31 were considered to have a good sample size, and 18 had a clear metric for effectiveness. **Transparency (T):** 44 studies included were published and peer-reviewed, of which 1 were literature reviews or meta-analyses, 0 were from the grey literature, and 0 were anecdotal. Of the studies

included, 30 had a published methodology, and 22 justified their rationale.

Provide supplementary food during the breeding season
Relevance (R): 2 studies in the evidence base focus on petrels and shearwaters, 14 on other seabirds and 0 on other birds. **Strength (S):** The evidence base was comprised of 16 studies. Of these 10 were considered to have a good sample size, and 14 had a clear metric for effectiveness. **Transparency (T):** 16 studies included were published and peer-reviewed, 0 were from the grey literature, and 0 were anecdotal. Of the studies included, 13 had a published methodology, and 4 justified their rationale.

Translocate the population to a more suitable breeding area
Relevance (R): 7 studies in the evidence base focus on petrels and shearwaters, 8 on other seabirds and 0 on other birds. **Strength (S):** The evidence base was comprised of 15 studies. Of these 13 were considered to have a good sample size, and 9 had a clear metric for effectiveness. **Transparency (T):** 14 studies included were published and peer-reviewed, of which 1 were literature reviews or meta-analyses, 0 were from the grey literature, and 0 were anecdotal. Of the studies included, 11 had a published methodology, and 9 justified their rationale.

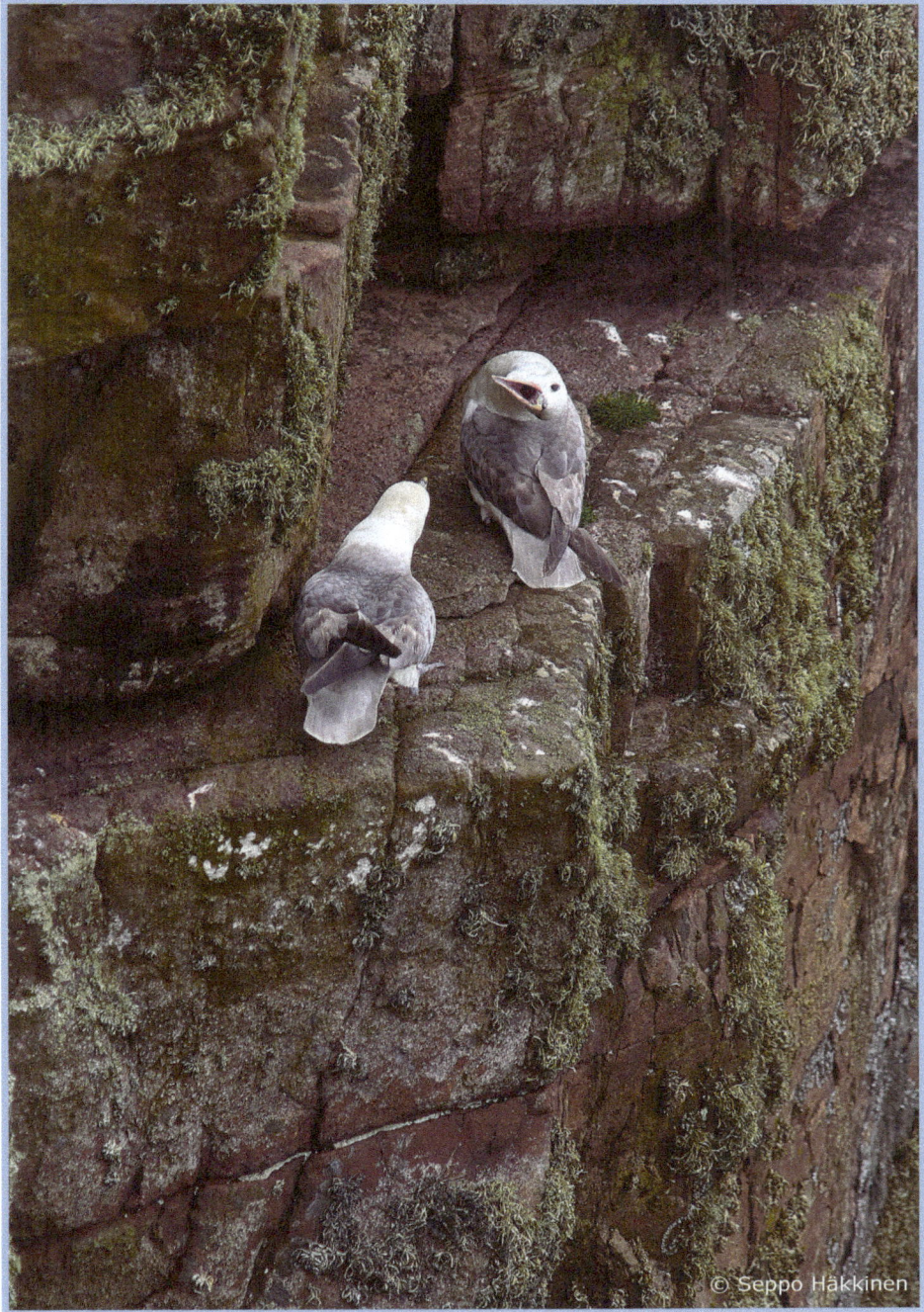

© Seppo Häkkinen

Northern fulmars

© Seppo Häkkinen

Skuas

(Stercorariidae)

An assessment of climate change vulnerability and potential conservation actions for skuas in the North-East Atlantic

UNIVERSITY OF
CAMBRIDGE

ZSL Institute
of Zoology

     https://doi.org/10.11647/OBP.0343.07

1 Great Skua (*Catharacta skua*)

1.1 Evidence for exposure

1.1.1 Potential changes in breeding habitat suitability (by 2100):

🟥 Current breeding area that is likely to become less suitable (78% of current range).

🟨 Current breeding area that is likely to remain suitable (19%).

🟩 Current breeding area that is likely to become more suitable (3%).

1.1.2 Current impacts attributed to climate change:

① Negative Impact: Hotter summers result in increased heat stress in adults and chicks. Adults more frequently leave nests unattended to thermoregulate, which exacerbates chick heat stress.

② Negative Impact: In hotter summers, adults more frequently leave nests unattended due to prey shortages and to thermoregulate, which results in higher chick mortality due to predation.

③ Negative Impact: Changes in prey availability during the breeding season have led to decreased fledgling success.

④ Positive Impact: Changes in prey availability have led to increased population size.

1.1.3 Predicted changes in key prey species:

No key prey species are predicted to decline for this species.

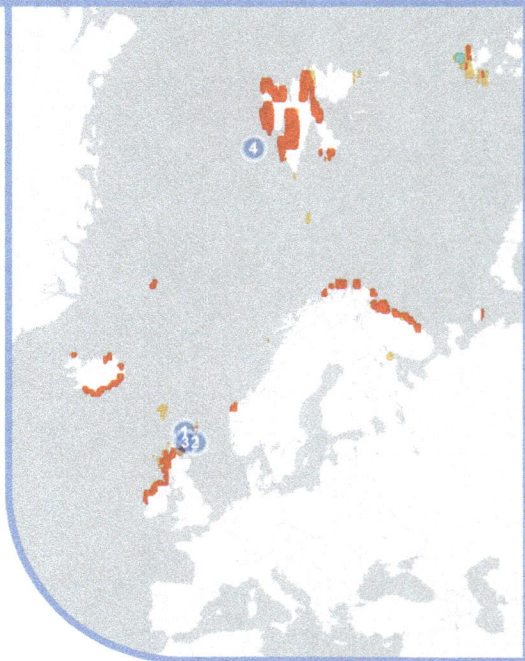

1.2 Sensitivity

•   Skuas are sensitive to high temperatures, and their southern range limit is likely defined by maximum temperature. Climate change is likely to make the southernmost populations unviable in the future.

•   Parasitism and predation of seabirds is an important part of skua diets, and climate change may heavily impact their prey species. In addition, they often cannibalise their neighbours, and this behaviour may increase as alternative prey becomes scarce.

•   This species has a long generation length (>10 years), which may slow recovery from severe impacts and increases population extinction risk.

1.3 Adaptive capacity

•   Great skuas have very varied diets and foraging strategies and will change their diet depending on availability. This flexibility may mean skuas can mitigate the impact of losing key prey species.

•   Great skuas are able to establish and colonise new areas, and have already done so at the northern edge of their range. They may be able to shift their range in response to climate change.

Arctic jaeger © Seppo Häkkinen

2 Long-tailed Jaeger

(*Stercorarius longicaudus*)

1.1 Evidence for exposure

1.1.1 Potential changes in breeding habitat suitability (by 2100):

■ Current breeding area that is likely to become less suitable (95% of current range).

■ Current breeding area that is likely to remain suitable (3%).

■ Current breeding area that is likely to become more suitable (2%).

1.1.2 Current impacts attributed to climate change:

❶ Negative Impact: Southern populations are becoming less populous or going extinct in correlation with rising temperatures. Exact mechanism unknown, probably related to prey availability or heat stress.

1.1.3 Predicted changes in key prey species:

No key prey assessment was carried out for this species.

1.1.4 Climate change impacts outside of Europe:

Long-tailed jaegers have been heavily affected by climate change in Greenland, in particular due to lack of prey and increased predation due to other species prey-switching.

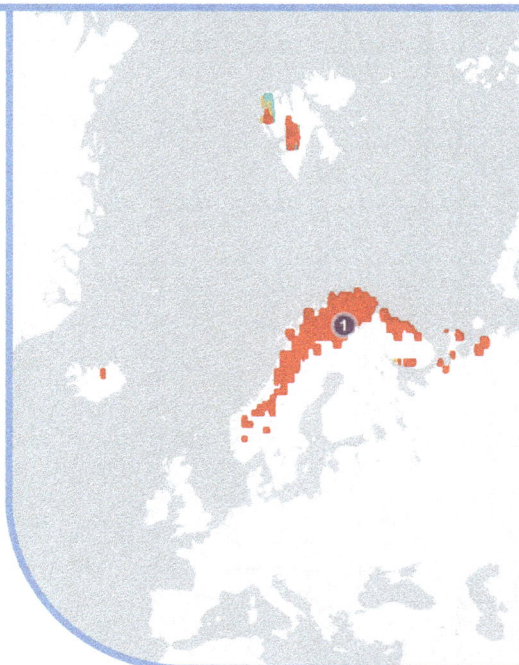

1.2 Sensitivity

• During the breeding season jaegers are heavily reliant on a few species of lemmings and voles, and any impact to these species is likely to heavily affect skua breeding success.

• Long-tailed jaeger populations are highly concentrated in the non-breeding season. >50% of global population congregate during migration in a relatively small area of the central Atlantic. Any negative change to this area is likely to have severe consequences on skua populations.

• Long-tailed jaeger chicks are highly vulnerable to predation by Arctic and red foxes (leading to up to 100% mortality in some years). Any changes in fox abundance (either negative or positive) may have severe impacts on long-tailed jaeger populations.

1.3 Adaptive capacity

• Jaegers are very site-tenacious so any response to change is likely to be very slow, and range shifts in the short term are very unlikely.

• Long-tailed jaegers will skip breeding in years with poor prey availability, which may be adaptive and maximises breeding output over time and help them cope with climate change. Long-tailed jaegers are long-lived and several years of breeding failure or skipped breeding may not have a long-term impact on the population if populations are able to breed successfully in good years.

Arctic jaeger © Seppo Häkkinen

3 Arctic Jaeger (*Stercorarius parasiticus*)

1.1 Evidence for exposure

1.1.1 Potential changes in breeding habitat suitability (by 2100):

■ Current breeding area that is likely to become less suitable (81% of current range).

■ Current breeding area that is likely to remain suitable (13%).

■ Current breeding area that is likely to become more suitable (6%).

1.1.2 Current impacts attributed to climate change:

① **Negative Impact:** Changes in prey availability have led to declines in key seabird species that Arctic jaegers parasitise, thus leading to population declines.

② **Negative Impact:** Increased competition and predation from great skuas, due to an increasing population size and prey swapping.

1.1.3 Predicted changes in key prey species:

No key prey assessment was carried out for this species.

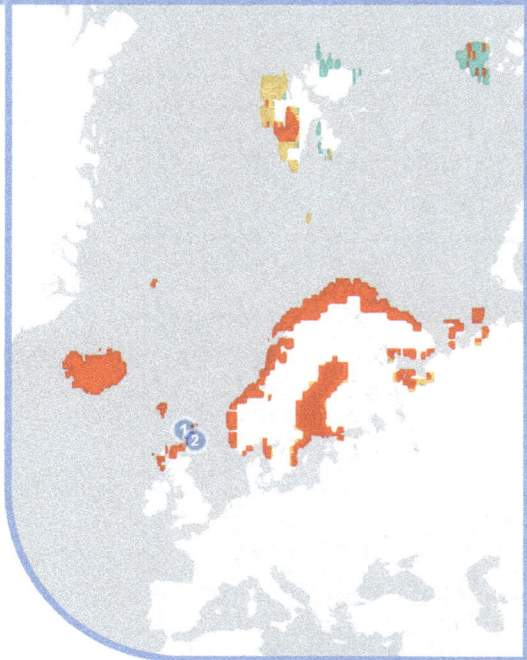

1.2 Sensitivity

• While Arctic jaegers parasitise a number of seabird species, the breeding success of many populations is closely linked the abundance of key fish species. Severe decreases in Arctic jaeger populations have been linked to prey declines, and changes in fish distributions due to climate change are likely to have heavy impacts on populations.

• Arctic jaeger populations are sensitive to predation, and several colonies have declined due to increased predation by great skuas and red foxes. Changes in predator abundance or range due to climate change (e.g. the expansion of red foxes in Scandinavia) are likely to have impacts on jaeger populations. In addition, jaeger skuas use co-operative defence which becomes less effective in smaller populations. This may result in a feedback loop where greater predation decreases population size, increasing vulnerability to predation.

• This species has a long generation length (>10 years), which may slow recovery from severe impacts and increases population extinction risk.

1.3 Adaptive capacity

• In most parts of their range Arctic jaegers are a numerous, long-lived, ecologically flexible species, so are likely to be robust to change.

• In some areas, particularly near major seabird colonies, Arctic jaegers have a quite restricted diet based on kleptoparasitism. However in many areas across their range they have a very varied diet, and will feed on the most available food, including birds, eggs, rodents, insects, fish, berries and carrion. This plasticity is likely to increase resilience to climate change, but variably across populations.

• There is considerable variation in migration route and wintering sites in Arctic jaegers, even within a single colony. This will likely provide a buffer to climate change, as changes to any one wintering site are less likely to affect the population as a whole.

Arctic jaeger © Seppo Häkkinen

Potential actions in response to climate change: Skuas (Stercorariidae)

In this section we list and assess possible local conservation actions that could be carried out in response to identified climate change impacts. This section is not grouped by species, but by identified impacts. If an impact or action is specific to one or a few species, this information is included in the action summary or in the footnotes. Effectiveness, relevance, strength and transparency scores are based on the available evidence we collated (see Appendix 2), and therefore all statements regarding limited or a lack of evidence relate to the collated evidence base, and does not infer that no such studies exist.

1 Impact: Increase in avian predation

Summary:

There are a number of available local actions to prevent or mitigate avian predation, some of which have been trialled extensively in seabirds with positive results. Other actions are poorly understood, but could be considered after more investigation. If predation is severe, and is likely to increase due to climate change or species range shifts, then translocation could be considered.

Intervention	Evidence of Effectiveness	R	S	T
Artificial shelters to make nests less visible to aerial predators	This is a hypothetical action. We found no published studies assessing this action's effectiveness for seabirds.	NA	NA	NA
Manage/ eradicate avian predators	Has been trialled with some success on several seabird groups, though has never been trialled for skua conservation. Often carried out as part of a suite of conservation actions, so difficult to assess how effective management is.	2	4	4

Physically protect nests with barriers or enclosures	Has been trialled on many seabird groups, often with notable success. Currently no reports on its effectiveness for skua conservation. A relatively easy, inexpensive method, but dependent on being able to access nest-sites and effectively protect them. As some skuas nest in very low densities across large areas of tundra, its practicality may be questionable.	2	4	4
Reduce predation by translocating predators	Few trials on seabirds, and none for skua conservation. Existing evidence suggests this action can be beneficial and reduce egg/chick predation, and could be a viable action if other forms of predator management are not viable.	1	4	3
Repel predators with acoustic, chemical or visual deterrents	This is a hypothetical action. We found no published studies assessing this action's effectiveness for seabirds.	NA	NA	NA
Use supplementary feeding to reduce predation	Very few trials on seabirds, and none for skua conservation. Likely to be very labour intensive and difficult given the remote and inaccessible breeding colonies of many skuas. More work is needed to examine action's effectiveness on seabirds.	1	4	3

Green = Likely to be beneficial. Red = Unlikely to be beneficial, may have negative impact.
Orange = contradicting or uncertain evidence. Grey = Limited evidence.
R = relevance rating. S = strength rating. T = transparency rating. All ratings on a scale of 1 to 5, where 5 is the highest.

Details:

Manage/eradicate avian predators
Relevance (R): 0 studies in the evidence base focus on skuas, 14 on other seabirds

and 2 on other birds. **Strength (S):** The evidence base was comprised of 16 studies. Of these 15 were considered to have a good sample size, and 5 had a clear metric for effectiveness. **Transparency (T):** 16 studies included were published and peer-reviewed, of which 1 were literature reviews or meta-analyses, 0 were from the grey literature, and 0 were anecdotal. Of the studies included, 9 had a published methodology, and 11 justified their rationale.

Physically protect nests with barriers or enclosures

Relevance (R): 0 studies in the evidence base focus on skuas, 12 on other seabirds and 6 on other birds. **Strength (S):** The evidence base was comprised of 18 studies. Of these 16 were considered to have a good sample size, and 12 had a clear metric for effectiveness. **Transparency (T):** 17 studies included were published and peer-reviewed, 0 were from the grey literature, and 0 were anecdotal. Of the studies included, 11 had a published methodology, and 12 justified their rationale.

Reduce predation by translocating predators

Relevance (R): 0 studies in the evidence base focus on skuas, 2 on other seabirds and 2 on other birds. **Strength (S):** The evidence base was comprised of 4 studies. Of these 4 were considered to have a good sample size, and 3 had a clear metric for effectiveness. **Transparency (T):** 4 studies included were published and peer-reviewed, 0 were from the grey literature, and 0 were anecdotal. Of the studies included, 2 had a published methodology, and 3 justified their rationale.

Use supplementary feeding to reduce predation

Relevance (R): 0 studies in the evidence base focus on skuas, 1 on other seabirds and 3 on other birds. **Strength (S):** The evidence base was comprised of 4 studies. Of these 4 were considered to have a good sample size, and 4 had a clear metric for effectiveness. **Transparency (T):** 4 studies included were published and peer-reviewed, 0 were from the grey literature, and 0 were anecdotal. Of the studies included, 1 had a published methodology, and 4 justified their rationale.

2 Impact: Increase in competition

Summary:
Local actions to prevent or mitigate the effects of competition are not well understood, and their effectiveness is unclear. In many contexts they are likely to be difficult or impossible to carry out on large populations. Supporting populations more generally (increasing adult survival, limiting chick mortality) may be a more appropriate strategy.

Intervention	Evidence of Effectiveness	R	S	T
Protect nest sites from competitors	Only trialled on one population of petrels (with limited success), all other examples focus on non-seabird species (many of which were successful). More work is needed to examine action's effectiveness on seabirds.	1	3	2
Reduce competition by removing competitor species	Trialled extensively on terns, but limited trials for other seabird groups, and none for skua conservation. Success is mixed, some trials have found benefits, but many have reported no effect or even negative consequences of this action.	2	3	3
Use supplementary feeding to reduce competition	This is a hypothetical action. We found no published studies assessing this action's effectiveness for seabirds.	NA	NA	NA

Green = Likely to be beneficial. Red = Unlikely to be beneficial, may have negative impact. Orange = contradicting or uncertain evidence. Grey = Limited evidence.
R = relevance rating. S = strength rating. T = transparency rating. All ratings on a scale of 1 to 5, where 5 is the highest.

Details:

Protect nest sites from competitors
Relevance (R): 0 studies in the evidence base focus on skuas, 2 on other seabirds and 5 on other birds. **Strength (S):** The evidence base was comprised of 7 studies. Of these 5 were considered to have a good sample size, and 2 had a clear metric for effectiveness. **Transparency (T):** 6 studies included were published and peer-reviewed, 0 were from the grey literature, and 0 were anecdotal. Of the studies included, 3 had a published methodology, and 4 justified their rationale.

Reduce competition by removing competitor species
Relevance (R): 0 studies in the evidence base focus on skuas, 12 on other seabirds and 0 on other birds. **Strength (S):** The evidence base was comprised of 12 studies. Of these 10 were considered to have a good sample size, and 5 had a clear metric for effectiveness. **Transparency (T):** 12 studies included were published and peer-reviewed, 0 were from the grey literature, and 0 were anecdotal. Of the studies included, 8 had a published methodology, and 7 justified their rationale.

3 Impact: Increased thermal stress

Summary:

There are currently no well-researched methods to directly assist seabirds with thermal stress, and more information is needed on how thermal stress can impact seabirds and how local conservation action can mitigate these impacts. If thermal stress becomes so common or extreme that it threatens the viability of a population, then several actions are available to encourage translocation of populations to safer areas.

Intervention	Evidence of Effectiveness	R	S	T
Make new colonies more attractive to encourage birds to colonise	Has been tried extensively on many different seabird groups with frequent, though not universal, success. However, currently there are no reports on this action's effectiveness for skuas.	3	4	3
Provide additional resources to help seabirds thermoregulate (e.g. artificial pools)	This is a hypothetical action. We found no published studies assessing this action's effectiveness for seabirds.	NA	NA	NA
Provide additional shelter or protection from extreme weather (heatwaves)	Very limited number of trials in seabirds, some limited benefits found for providing additional shelter from the sun for cormorants. More work is needed to examine action's effectiveness on seabirds.	2	3	3
Translocate the population to a more suitable breeding area	Known to be beneficial in some seabird groups, but no recorded trials in skuas. Skuas tend to have extremely high territoriality and site-fidelity so translocation of adults is likely to be extremely difficult, if not impossible. Whether translocation is plausible, or beneficial, to skuas is currently unknown and further research is needed.	3	4	4

Details:

Make new colonies more attractive to encourage birds to colonise

Relevance (R): 0 studies in the evidence base focus on skuas, 38 on other seabirds and 6 on other birds. **Strength (S):** The evidence base was comprised of 44 studies. Of these 31 were considered to have a good sample size, and 18 had a clear metric for effectiveness. **Transparency (T):** 44 studies included were published and peer-reviewed, of which 1 were literature reviews or meta-analyses, 0 were from the grey literature, and 0 were anecdotal. Of the studies included, 30 had a published methodology, and 22 justified their rationale.

Provide additional shelter or protection from extreme weather (heatwaves)

Relevance (R): 0 studies in the evidence base focus on skuas, 1 on other seabirds and 0 on other birds. **Strength (S):** The evidence base was comprised of 1 study. Of these 1 was considered to have a good sample size, and 1 had a clear metric for effectiveness. **Transparency (T):** 1 study included were published and peer-reviewed, 0 were from the grey literature, and 0 were anecdotal. Of the studies included, 0 had a published methodology, and 1 justified their rationale.

Translocate the population to a more suitable breeding area

Relevance (R): 0 studies in the evidence base focus on skuas, 15 on other seabirds and 0 on other birds. **Strength (S):** The evidence base was comprised of 15 studies. Of these 13 were considered to have a good sample size, and 9 had a clear metric for effectiveness. **Transparency (T):** 14 studies included were published and peer-reviewed, of which 1 were literature reviews or meta-analyses, 0 were from the grey literature, and 0 were anecdotal. Of the studies included, 11 had a published methodology, and 9 justified their rationale.

4 Impact: Reduced prey availability during breeding season

Summary:
Several local actions may assist breeding populations on a small scale, but direct intervention on a large scale is likely to be extremely difficult. General conservation actions to protect fish stocks and local marine areas may be the most effective method. If a population is likely to suffer major losses, even with conservation help, then translocations could be considered.

Intervention	Evidence of Effectiveness	R	S	T
Artificially incubate or hand-rear chicks to support population	Known to be effective for some seabirds, though labour intensive and usually only appropriate for small populations. To our knowledge, there are no examples of skuas being hand-reared successfully, though there are reports of previous ex-situ populations.	2	2	1
Make new colonies more attractive to encourage birds to colonise	Has been tried extensively on many different seabird groups with frequent, though not universal, success. However, currently there are no reports on this action's effectiveness for skuas.	2	4	3
Provide supplementary food during the breeding season	Trialled on several seabird species, with some, though not universal, success. Trialled on only one population of skuas, which found little benefit. Typically very labour intensive and difficult given the remote and inaccessible breeding colonies of many skuas. Probably only plausible for small populations.	3	4	3
Translocate the population to a more suitable breeding area	Known to be beneficial in some seabird groups, but no recorded trials in skuas. Skuas tend to have extremely high territoriality and site-fidelity so translocation of adults is likely to be extremely difficult, if not impossible. Whether translocation is plausible, or beneficial, to skuas is currently unknown and further research is needed.	2	4	4

Green = Likely to be beneficial. Red = Unlikely to be beneficial, may have negative impact. Orange = contradicting or uncertain evidence. Grey = Limited evidence.
R = relevance rating. S = strength rating. T = transparency rating. All ratings on a scale of 1 to 5, where 5 is the highest.

Details:

Artificially incubate or hand-rear chicks to support population
Relevance (R): 0 studies in the evidence base focus on skuas, 40 on other seabirds and 0 on other birds. **Strength (S):** The evidence base was comprised of 40 studies. Of these 9 were considered to have a good sample size, and 19 had a clear metric for effectiveness. **Transparency (T):** 26 studies included were published and peer-reviewed, 0 were from the grey literature, and 0 were anecdotal. Of the studies included, 17 had a published methodology, and 4 justified their rationale.

Make new colonies more attractive to encourage birds to colonise
Relevance (R): 0 studies in the evidence base focus on skuas, 38 on other seabirds and 6 on other birds. **Strength (S):** The evidence base was comprised of 44 studies. Of these 31 were considered to have a good sample size, and 18 had a clear metric for effectiveness. **Transparency (T):** 44 studies included were published and peer-reviewed, of which 1 were literature reviews or meta-analyses, 0 were from the grey literature, and 0 were anecdotal. Of the studies included, 30 had a published methodology, and 22 justified their rationale.

Provide supplementary food during the breeding season
Relevance (R): 2 studies in the evidence base focus on skuas, 14 on other seabirds and 0 on other birds. **Strength (S):** The evidence base was comprised of 16 studies. Of these 10 were considered to have a good sample size, and 14 had a clear metric for effectiveness. **Transparency (T):** 16 studies included were published and peer-reviewed, 0 were from the grey literature, and 0 were anecdotal. Of the studies included, 13 had a published methodology, and 4 justified their rationale.

Translocate the population to a more suitable breeding area
Relevance (R): 0 studies in the evidence base focus on skuas, 15 on other seabirds and 0 on other birds. **Strength (S):** The evidence base was comprised of 15 studies. Of these 13 were considered to have a good sample size, and 9 had a clear metric for effectiveness. **Transparency (T):** 14 studies included were published and peer-reviewed, of which 1 were literature reviews or meta-analyses, 0 were from the grey literature, and 0 were anecdotal. Of the studies included, 11 had a published methodology, and 9 justified their rationale.

© Seppo Häkkinen

Terns
(Laridae)

An assessment of climate change vulnerability and potential conservation actions for terns in the North-East Atlantic

UNIVERSITY OF
CAMBRIDGE

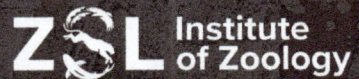

ZSL Institute
of Zoology

    https://doi.org/10.11647/OBP.0343.08

1 Arctic Tern *(Sterna paradisaea)*

1.1 Evidence for exposure

1.1.1 Potential changes in breeding habitat suitability (by 2100):

🟥 Current breeding area that is likely to become less suitable (87% of current range).

🟨 Current breeding area that is likely to remain suitable (10%).

🟩 Current breeding area that is likely to become more suitable (3%).

1.1.2 Current impacts attributed to climate change:

① **Negative Impact:** Changes in prey availability during the breeding season have led to population declines (debated).

② **Negative Impact:** Changes in prey availability during the breeding season have led to population declines.

③ **Neutral Impact:** Arctic terns are arriving from migration and breeding earlier.

④ **Neutral Impact:** Juvenile Arctic terns have begun to disperse further, distance has increased in correlation with warmer winters. Mechanism unknown, but likely mediated through prey availability.

1.1.3 Predicted changes in key prey species:

⑤ Key prey species are likely to decline in abundance in the Irish Sea and around North Denmark.

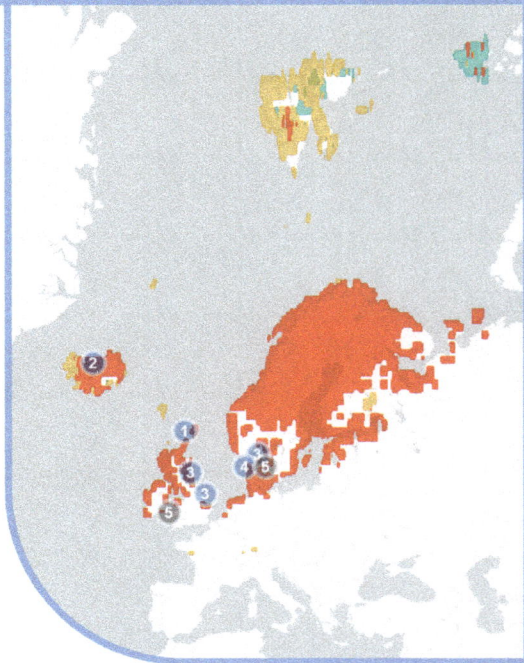

1.2 Sensitivity

• Around the North Sea Arctic terns are very dependent on sandeels and highly sensitive to prey depletion. Lack of sandeels can cause mass breeding failures, and several have been observed in recent years. If climate change contributes to the decline of sandeels this is likely to have severe consequences on tern populations.

• Tern productivity decreases as sea surface temperature gets higher, most likely due to prey availability. While the mechanism is unclear, projected increases in sea temperature are likely to negatively impact tern populations.

• Arctic terns are primarily surface feeders. If climate change results in prolonged stormy weather, or extended heatwaves drive prey species into deeper water, then it would likely result in Arctic terns struggling to forage effectively.

• This species has a long generation length (>10 years), which may slow recovery from severe impacts and increases population extinction risk.

1.3 Adaptive capacity

• Dispersal between colonies in Denmark has steadily increased over the last 70 years, in correlation with changes in climate (winter NAO). The exact mechanism is unknown, but high dispersal typically means more genetic flow and therefore resilience in the population.

• Arctic terns can skip breeding when conditions are particularly poor. This is likely to be adaptive in the face of climate change as it conserves resources in poor years and maximises breeding success in good years

• Arctic tern populations have a wide variation in migration strategy in terms of route and timing. This is likely to make populations more resilient as impacts to one part of the migration route are unlikely to have major impacts on the population as a whole.

• Arctic terns likely have high site fidelity, particularly at stable colonies. There are few examples of Arctic terns establishing new colonies spontaneously. They are unlikely to shift their range quickly in response to climate change.

2 Little Tern *(Sternula albifrons)*

1.1 Evidence for exposure

1.1.1 Potential changes in breeding habitat suitability (by 2100):

■ Current breeding area that is likely to become less suitable (3% of current range).

■ Current breeding area that is likely to remain suitable (57%).

■ Current breeding area that is likely to become more suitable (59%).

1.1.2 Current impacts attributed to climate change:

① **Negative Impact:** Little tern nests are frequently washed away by tidal surges, such events are becoming more frequent or extensive due to rising sea levels.

② **Negative Impact:** As sea temperature has increased over time, tern productivity has decreased. Mechanism unknown, but likely mediated through prey availability.

1.1.3 Predicted changes in key prey species:

③ Key prey species are likely to decline in abundance in southern Portugal, along the southern coast of England, the coasts of Belgium and the Netherlands and around North Denmark.

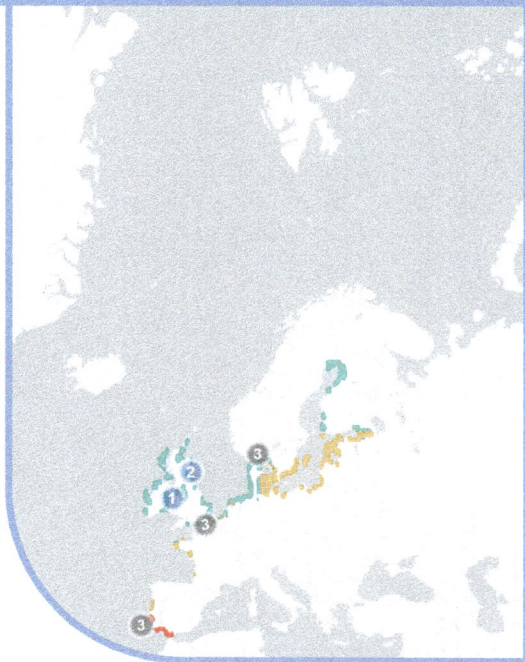

1.2 Sensitivity

- Little terns have a varied diet, but many colonies rely on one or a few prey species. Sensitivity to changes in prey is likely to vary across its range. In some populations (such as in Algarve, Portugal), warmer temperatures have been linked to lower key prey availability and lower breeding success, though there has been no long-term trend observed over time.
- Coastal populations breed commonly on flat beaches, which are prone to flooding due to storms, tidal surges or sea level rise. A rise in extreme weather events would likely impact breeding success of little terns.
- Little terns are primarily surface feeders and have a limited foraging range during the breeding season. This increases their sensitivity to changes in prey availability, and they would likely be heavily impacted if climate change results in prolonged stormy weather, or extended heatwaves drive prey species into deeper water.
- Little terns face high levels of predation threat, notably in the Baltic, and several of their key predators (e.g. minks and foxes) are becoming more abundant and spreading, in part due to climate change. So far this has not heavily impacted little tern populations, but continued climate change may result in declines due to predation.

1.3 Adaptive capacity

- Little terns can change their phenology based on climate and weather. Populations in Finland change their migration timing in response to winter and spring climate, and their laying date in response to local weather conditions. This may allow little terns to respond to changing climate and mitigate impacts leading up to the breeding season.
- Little terns often display low site fidelity and will change breeding site from year to year, especially in response to disturbance. This is likely to be adaptive in terms of climate change, as it seems highly likely little terns will redistribute to more suitable areas if available.

3 Roseate Tern *(Sterna dougallii)*

1.1 Evidence for exposure

1.1.1 Potential changes in breeding habitat suitability (by 2100):

■ Current breeding area that is likely to become less suitable (100% of current range).

■ Current breeding area that is likely to remain suitable (0%).

■ Current breeding area that is likely to become more suitable (0%).

1.1.2 Current impacts attributed to climate change:

We did not identify any current impacts of climate change for this species.

1.1.3 Predicted changes in key prey species:

① Key prey species are likely to decline in abundance in the English Channel.

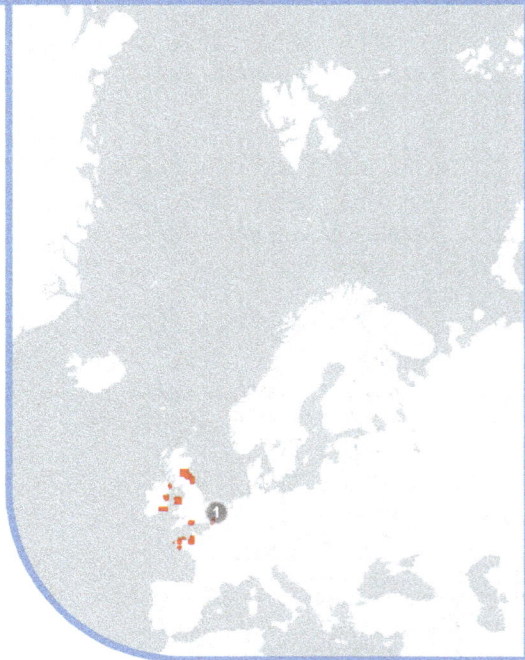

1.2 Sensitivity

• Roseate terns have a small range in Europe, and many populations have historically gone extinct due to persecution, disturbance, or extreme climate events. Increased pressure from climate change is likely to increase risk of declines or local extinction.

• Roseate terns are primarily surface feeders. If climate change results in prolonged stormy weather, or extended heatwaves drive prey species into deeper water, then it would likely result in terns struggling to forage effectively.

• Roseate terns are specialist foragers, and prey on relatively few species compared to other terns. They may be particularly sensitive to changes in key prey species.

• Many European Roseate terns winter off the coast of west Africa, where prey availability is linked to key upwelling systems. There is evidence that these upwellings may change or be disrupted by climate change. If this happens it is likely to have severe consequences on tern mortality, condition during the breeding season and timing of migration.

• Roseate terns only nest around other terns, in particular common terns, and rely on them for protection from more aggressive species. Any impacts to these other species is likely to have negative consequences for roseate terns.

1.3 Adaptive capacity

• The eastern Atlantic roseate tern population is thought to be a closed population, so is unlikely to be supplemented by individuals from other colonies. Genetic diversity is unlikely to increase and may result in bottlenecks, possibly hampering recovery if climate change results in population decline.

Arctic tern © Seppo Häkkinen

4 Sandwich Tern *(Thalasseus sandvicensis)*

1.1 Evidence for exposure

1.1.1 Potential changes in breeding habitat suitability (by 2100):

■ Current breeding area that is likely to become less suitable (72% of current range).

■ Current breeding area that is likely to remain suitable (26%).

■ Current breeding area that is likely to become more suitable (2%).

1.1.2 Current impacts attributed to climate change:

① **Neutral Impact:** Sandwich terns are changing their migration timing and arriving earlier to breeding sites.

② **Neutral Impact:** Sandwich terns are changing their migration timing, both migration and breeding events are occurring later, making the breeding season shorter.

1.1.3 Predicted changes in key prey species:

③ Key prey species are likely to decline in abundance on the southern coast of England, around the Brittany coast, around North Denmark and around the south coast of Ireland.

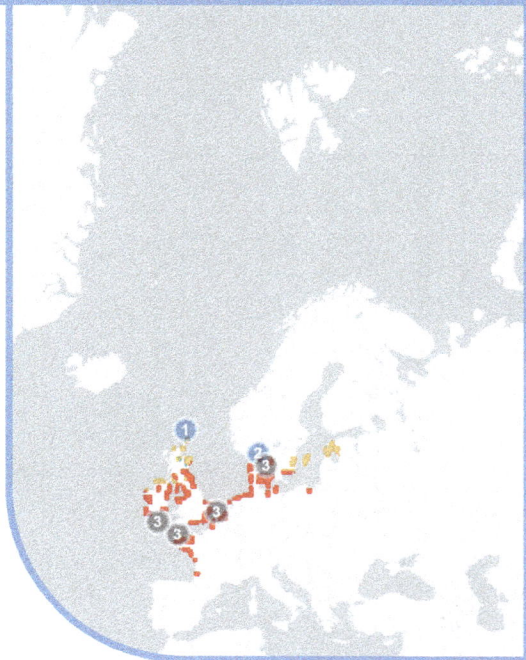

1.2 Sensitivity

• As Sandwich terns nest on low-lying ground close to the water's edge, their nests are vulnerable to tidal inundation. So far there are no studies on whether climate change is affecting breeding populations through increased flooding, but predictions of increased storminess and sea-level rise under climate change scenarios may lead to more nesting populations being lost.

1.3 Adaptive capacity

• Sandwich terns have flexible migration and laying phenology, and several populations have already changed the timing of migration and breeding in response to climate change. However, whether this is adaptive or not is currently unknown.

• Sandwich terns often display low site fidelity and will change breeding site from year to year in response to changing conditions or disturbance. This is likely to be adaptive in terms of climate change, as it seems highly likely Sandwich terns will redistribute to more suitable areas if available.

Sandwich tern © Silviu Petrovan

5 Caspian Tern *(Hydroprogne caspia)*

1.1 Evidence for exposure

1.1.1 Potential changes in breeding habitat suitability (by 2100):

■ Current breeding area that is likely to become less suitable (97% of current range).

■ Current breeding area that is likely to remain suitable (3%).

■ Current breeding area that is likely to become more suitable (0%).

1.1.2 Current impacts attributed to climate change:

We did not identify any current impacts of climate change for this species.

1.1.3 Predicted changes in key prey species:

No key prey species are predicted to decline for this species.

1.1.4 Climate change impacts outside of Europe

• Caspian terns in North America have been negatively affected by heatwaves, warming seas, severe storms, and increased frequency of flooding, all of which are linked to climate change.

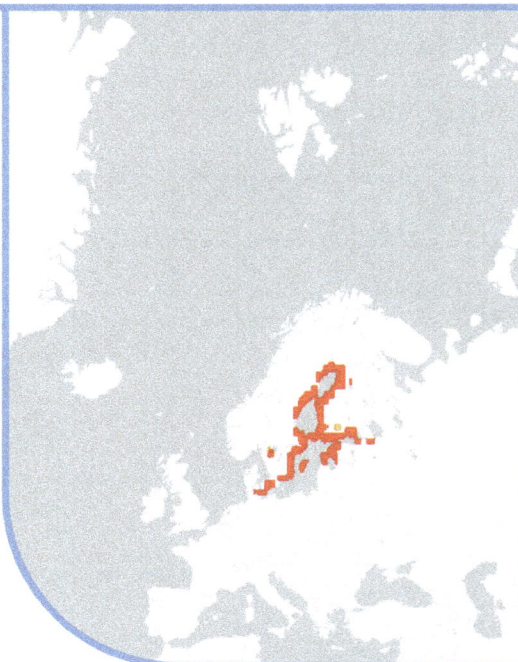

1.2 Sensitivity

- Caspian tern chicks are sensitive to heatwaves, and particularly hot summers in the US have resulted in mass mortality. Heatwaves are likely to become more frequent and extreme due to climate change, which will likely negatively impact Caspian terns.
- Caspian terns are vulnerable to predation by American minks, which are increasing in some locations in part due to climate change. No effect on populations has been observed so far, but climate change may contribute to additional predation pressure.
- Caspian terns in Europe typically nest on low-lying beaches, making them sensitive to flooding and tidal surges. Extreme weather events are projected to become more frequent in many areas, and therefore may lower tern breeding success.
- Caspian terns are primarily surface feeders. If climate change results in prolonged stormy weather, or extended heatwaves drive prey species into deeper water, then it would likely result in Caspian terns struggling to forage effectively.
- Caspian terns typically only nest on rocky substrates with very little vegetation. Additional growth can cause them to abandon nesting sites. This is particularly an issue in the Baltic due to high levels of nutrients on islands, and is likely to be exacerbated by climate change.

1.3 Adaptive capacity

- Caspian terns are known to abandon and re-locate colonies following changes in environmental conditions or disturbance. This is likely to be adaptive in terms of climate change, as it seems highly likely terns will redistribute to more suitable areas if available.

Potential actions in response to climate change: Terns (Laridae)

In this section we list and assess possible local conservation actions that could be carried out in response to identified climate change impacts. This section is not grouped by species, but by identified impacts. If an impact or action is specific to one or a few species, this information is included in the action summary or in the footnotes. Effectiveness, relevance, strength and transparency scores are based on the available evidence we collated (see Appendix 2), and therefore all statements regarding limited or a lack of evidence relate to the collated evidence base, and does not infer that no such studies exist.

1 Impact: Reduced prey availability during breeding season

Summary:

Several local actions may assist breeding populations on a small scale, but direct intervention on a large scale is likely to be extremely difficult. General conservation actions to protect fish stocks and local marine areas may be the most effective method. If a population is likely to suffer major losses, even with conservation help, then translocations could be considered.

Intervention	Evidence of Effectiveness	R	S	T
Artificially incubate or hand-rear chicks to support population	Known to be effective for some seabirds, though labour intensive and usually only appropriate for small populations. There are anecdotal examples of terns being hand-reared, but details are unknown. Arctic terns have been kept ex-situ in small numbers.	2	2	1
Make new colonies more attractive to encourage birds to colonise	Many species of tern have been successfully relocated using a variety of techniques to make new areas more attractive, in particular vocalisations and modification of habitat. While not every attempted relocation has been successful, there are numerous notable successes, including successful relocation of large populations containing thousands of individuals.	4	4	3

		R	S	T
Provide supplementary food during the breeding season	Trialled on many seabird species, with mixed success. Few, if any, successful examples for terns, though some supplemental feeding has been attempted on a very local scale. Typically very labour intensive and difficult, especially as many tern species strongly prefer living prey and are unlikely to take dead prey. Unlikely to be effective for many terns, or only plausible for small populations.	2	4	3
Translocate the population to a more suitable breeding area	Known to be beneficial in other seabird groups, but few if any examples in terns. Attempts to move populations have typically used attractants to encourage movement, rather than manual translocation. If the target species has high site fidelity, manual translocation may be appropriate but little is known about methods and effectiveness of tern translocation.	2	4	4

Green = Likely to be beneficial. Red = Unlikely to be beneficial, may have negative impact.
Orange = contradicting or uncertain evidence. Grey = Limited evidence.
R = relevance rating. S = strength rating. T = transparency rating. All ratings on a scale of 1 to 5, where 5 is the highest.

Details:

Artificially incubate or hand-rear chicks to support population

Relevance (R): 0 studies in the evidence base focus on terns, 40 on other seabirds and 0 on other birds. Strength (S): The evidence base was comprised of 40 studies. Of these 9 were considered to have a good sample size, and 19 had a clear metric for effectiveness. Transparency (T): 26 studies included were published and peer-reviewed, 0 were from the grey literature, and 0 were anecdotal. Of the studies included, 17 had a published methodology, and 4 justified their rationale.

Make new colonies more attractive to encourage birds to colonise

Relevance (R): 22 studies in the evidence base focus on terns, 16 on other seabirds and 6 on other birds. Strength (S): The evidence base was comprised of 44 studies. Of these 31 were considered to have a good sample size, and 18 had a clear metric for effectiveness. Transparency (T): 44 studies included were published and peer-reviewed, of which 1 were literature reviews or meta-analyses, 0 were from the grey literature, and 0 were anecdotal. Of the studies included, 30 had a published methodology, and 22 justified their rationale.

Provide supplementary food during the breeding season
Relevance (R): 0 studies in the evidence base focus on terns, 16 on other seabirds and 0 on other birds. **Strength (S):** The evidence base was comprised of 16 studies. Of these 10 were considered to have a good sample size, and 14 had a clear metric for effectiveness. **Transparency (T):** 16 studies included were published and peer-reviewed, 0 were from the grey literature, and 0 were anecdotal. Of the studies included, 13 had a published methodology, and 4 justified their rationale.

Translocate the population to a more suitable breeding area
Relevance (R): 0 studies in the evidence base focus on terns, 15 on other seabirds and 0 on other birds. **Strength (S):** The evidence base was comprised of 15 studies. Of these 13 were considered to have a good sample size, and 9 had a clear metric for effectiveness. **Transparency (T):** 14 studies included were published and peer-reviewed, of which 1 were literature reviews or meta-analyses, 0 were from the grey literature, and 0 were anecdotal. Of the studies included, 11 had a published methodology, and 9 justified their rationale.

2 Impact: Increased frequency/severity of storms (including wind, rain and wave action) causes nest destruction

Summary:
While there are several local actions that may prevent or mitigate local nest destruction, they have not been trialled widely and wide-spread evidence to support their use is currently lacking. If changes in extreme weather threatens the viability of a population, then several actions are available to encourage translocation of populations to safer areas.

Intervention	Evidence of Effectiveness	R	S	T
Alter habitat to encourage birds to leave an area	There are limited trials of this action, however there are several successful examples of modifying habitat to encourage terns to leave, typically by encouraging vegetation to overgrow nesting areas. There are currently no available examples in other seabirds.	5	2	3

Artificially incubate or hand-rear chicks to support population	Known to be effective for some seabirds, though labour intensive and usually only appropriate for small populations. There are anecdotal examples of terns being hand-reared, but details are unknown. Arctic terns have been kept ex-situ in small numbers.	2	2	1
Install barriers to prevent flooding	This is a hypothetical action. We found no published studies assessing this action's effectiveness for seabirds.	NA	NA	NA
Make new colonies more attractive to encourage birds to colonise	Many species of tern have been successfully relocated using a variety of techniques to make new areas more attractive, in particular vocalisations and modification of habitat. While not every attempted relocation has been successful, there are numerous notable successes, including successful relocation of large populations containing thousands of individuals.	4	4	3
Manually relocate nests	There is no published evidence regarding this action's effectiveness. However, first-hand accounts from practitioners have reported that manually moving nests has been trialled, and with some success. Likely to only be viable for a few nests at the edge of colonies, as larger interventions are laborious, cause mass disturbance, and may result in terns abandoning nests.	NA	NA	NA
Provide additional shelter or protection from extreme weather (flooding)	There have been some attempts to provide shelter for tern populations to protect nests from flooding, however none so far have shown any significant benefit. Novel methods may provide more protection, but this requires further research.	2	3	5

		R	S	T
Provide artificial nesting sites	Tried extensively on many seabird species with significant benefit to many species. Artificial nesting sites have been successfully used to support a variety of tern species, including the use of artificial islands, floating rafts and nest-boxes.	3	5	3
Repair/ support nests to support breeding	There is limited evidence to support this action, as few trials have been carried out on any seabird species, and those that exist are on a very local scale. No attempts have been made on terns. More research needed to determine effectiveness of this action.	2	2	3
Translocate the population to a more suitable breeding area	Known to be beneficial in other seabird groups, but few, if any, examples in terns. Attempts to move populations have typically used attractants to encourage movement, rather than manual translocation. If the target species has high site fidelity, manual translocation may be appropriate but little is known about methods and effectiveness of tern translocation.	4	4	4

Green = Likely to be beneficial. Red = Unlikely to be beneficial, may have negative impact. Orange = contradicting or uncertain evidence. Grey = Limited evidence.
R = relevance rating. S = strength rating. T = transparency rating. All ratings on a scale of 1 to 5, where 5 is the highest.

Details:

Alter habitat to encourage birds to leave an area
Relevance (R): 2 studies in the evidence base focus on terns, 0 on other seabirds and 0 on other birds. **Strength (S):** The evidence base was comprised of 2 studies. Of these 2 were considered to have a good sample size, and 0 had a clear metric for effectiveness. **Transparency (T):** 2 studies included were published and peer-reviewed, 0 were from the grey literature, and 0 were anecdotal. Of the studies included, 2 had a published methodology, and 1 justified their rationale.

Artificially incubate or hand-rear chicks to support population
Relevance (R): 0 studies in the evidence base focus on terns, 40 on other seabirds and 0 on other birds. **Strength (S):** The evidence base was comprised of 40 studies. Of these 9 were considered to have a good sample size, and 19 had a clear metric for effectiveness. **Transparency (T):** 26 studies included were published and peer-reviewed, 0 were from the grey literature, and 0 were anecdotal. Of the studies

included, 17 had a published methodology, and 4 justified their rationale.

Make new colonies more attractive to encourage birds to colonise
Relevance (R): 22 studies in the evidence base focus on terns, 16 on other seabirds and 6 on other birds. **Strength (S):** The evidence base was comprised of 44 studies. Of these 31 were considered to have a good sample size, and 18 had a clear metric for effectiveness. **Transparency (T):** 44 studies included were published and peer-reviewed, of which 1 were literature reviews or meta-analyses, 0 were from the grey literature, and 0 were anecdotal. Of the studies included, 30 had a published methodology, and 22 justified their rationale.

Provide additional shelter or protection from extreme weather (flooding)
Relevance (R): 1 study in the evidence base focusses on terns, 0 on other seabirds and 2 on other birds. **Strength (S):** The evidence base was comprised of 3 studies. Of these 1 was considered to have a good sample size, and 2 had a clear metric for effectiveness. **Transparency (T):** 3 studies included were published and peer-reviewed, 0 were from the grey literature, and 0 were anecdotal. Of the studies included, 3 had a published methodology, and 3 justified their rationale.

Provide artificial nesting sites
Relevance (R): 17 studies in the evidence base focus on terns, 35 on other seabirds and 1 on other birds. **Strength (S):** The evidence base was comprised of 54 studies. Of these 50 were considered to have a good sample size, and 33 had a clear metric for effectiveness. **Transparency (T):** 53 studies included were published and peer-reviewed, of which 2 were literature reviews or meta-analyses, 0 were from the grey literature, and 0 were anecdotal. Of the studies included, 33 had a published methodology, and 27 justified their rationale.

Repair/support nests to support breeding
Relevance (R): 1 study in the evidence base focusses on terns, 1 on other seabirds and 1 on other birds. **Strength (S):** The evidence base was comprised of 3 studies. Of these 1 was considered to have a good sample size, and 1 had a clear metric for effectiveness. **Transparency (T):** 3 studies included were published and peer-reviewed, 0 were from the grey literature, and 0 were anecdotal. Of the studies included, 1 had a published methodology, and 3 justified their rationale.

Translocate the population to a more suitable breeding area
Relevance (R): 0 studies in the evidence base focus on terns, 15 on other seabirds and 0 on other birds. Strength (S): The evidence base was comprised of 15 studies. Of these 13 were considered to have a good sample size, and 9 had a clear metric for effectiveness. Transparency (T): 14 studies included were published and peer-reviewed, of which 1 were literature reviews or meta-analyses, 0 were from the grey literature, and 0 were anecdotal. Of the studies included, 11 had a published methodology, and 9 justified their rationale.

© Seppo Häkkinen

Appendices

Appendix 1: Auks
Sources and references for vulnerability assessment

1.1 Evidence for exposure (references)

1.1.1 Current impacts attributed to climate change:

Razorbill

1 - Extreme storms during the razorbill breeding season have led to wide-spread nest destruction, nesting failure and a net reduction in annual population production

Newell, M., et al. "Effects of an extreme weather event on seabird breeding success at a North Sea colony." Marine Ecology Progress Series 532 (2015): 257-268. A single extreme summer storm on the Isle of May resulted in wide-spread nest destruction, nesting failure and a net reduction in annual population production. While individual storms cannot be easily be attributed to climate change, severe storms are increasingly frequent in Europe.

2 - As sea temperatures have increased over time, razorbill productivity has decreased, most likely due to changes in prey availability.

Lauria, V., et al. "Influence of climate change and trophic coupling across four trophic levels in the Celtic Sea." (2012): e47408. Razorbill productivity on Skomer Island declined over the study period (1993–2007). Meanwhile, spring sea surface temperature significantly increased. Razorbill productivity was correlated with spring sea surface temperature in the previous year. The study suggests that sea temperature affected the razorbills indirectly, through the availability of forage fish.

3 - Key prey species have shifted their life-cycle, likely in response to climate change, but razorbills have not adjusted in response. There is concern this could result in trophic mismatch, but no overall effect on breeding success has so far been observed.

Burthe, S., et al. "Phenological trends and trophic mismatch across multiple levels of a North Sea pelagic food web." Marine Ecology Progress Series 454 (2012): 119-133. Timing of sandeel growth has changed substantially, but laying date has not in razorbills. This likely has resulted in trophic mismatch. However, to date no overall effect on breeding success has been observed. Seabird observations based mostly on Isle of May.

Little Auk

1 - Warmer temperatures correlate with longer foraging trips and lower little auk productivity, most likely due to decreased prey availability.

Hovinen, J. E. H., et al. "Climate warming decreases the survival of the little auk (Alle alle), a high Arctic avian predator." Ecology and Evolution 4.15 (2014): 3127-3138. At several sites in Svalbard, higher SST is associated with decreased adult survival, probably mediated through prey availability. Suggested mechanism is that an increase in temperature results in a decrease in sea ice and a decrease in ice algal production which in turn results in less food quality and availability. The temperature and inflow of warm Atlantic water to the Arctic increased during the study period.

Hovinen, J. E. H., et al. "Fledging success of little auks in the high Arctic: do provisioning rates and the quality of foraging grounds matter?." Polar Biology 37.5 (2014): 665-674. Higher SST in colonies around Svalbard correlates with lower fledging success, though not with provisioning rate by parents. In at least one colony, this has led to population decline. Most likely linked to higher SST resulting in lower prey availability quality.

Ramírez, F., et al. "Sea ice phenology and primary productivity pulses shape breeding success in Arctic seabirds." Scientific Reports 7.1 (2017): 1-9. As above, warmer years with less sea ice result in changes in timing of key prey species availability. This in turn correlates with lower breeding performance in little auks. Study was based around Spitsbergen, Svalbard.

Jakubas, D., Wojczulanis-Jakubas, K., and Walkusz, W. "Response of dovekie to changes in food availability." Waterbirds 30.3 (2007): 421-428. This study looks at similar effects to those above, they note warmer waters around Spitsbergen means less easily accessible high-quality food, but that adults are able to compensate somewhat with changes in their foraging strategy. Study also notes that the North Atlantic has warmed over recent years.

2 - Little auks are breeding earlier in correlation with warmer temperatures, so far no negative consequence has been observed

Moe, Børge, et al. "Climate change and phenological responses of two seabird species breeding in the high-Arctic." Marine Ecology Progress Series 393 (2009): 235-246. Little auks on Svalbard are breeding earlier in correlation with increases in air temperature in the spring. The reason for this is not clear, but may be a result of nesting sites being available earlier due to snow melt. This change in phenology may or may not match prey availability, which may lead to trophic mismatch in the future.

3 - Extreme storms during the non-breeding season have led to mass mortality of

little auks ('wrecks')

> Clairbaux, M., et al. "North Atlantic winter cyclones starve seabirds."
> Current Biology 31.17 (2021): 3964-3971. Following heavy storm action,
> seabird mortality increases due to increased difficulty foraging (rather than
> increased energetic costs). The authors use a multi-species dataset (puffins,
> little auks, common murres, and thick-billed murres) over a wide area of the
> Atlantic basin. They conclude that seabirds around Iceland and the Barents Sea
> (along with several N. American sites) are particularly vulnerable. Climate
> change is likely to be a contributing factor to present and future storm mortality.

Black Guillemot

1 - Heavy rainfall events and high water level has led to flooding of nests and lower
hatching success in the Baltic. The authors note that such flooding events are likely
to further increase. Debris left by storms and flooding can also make large areas of
shoreline less suitable for breeding

> Hof, A. R., Crombag, J. A. H. M. , and Allen, A. M. "The ecology of Black
> Guillemot Cepphus grylle grylle chicks in the Baltic Sea region: insights into
> their diet, survival, nest predation and moment of fledging." Bird Study
> 65.3 (2018): 357-364. Storms and increased water level leads to flooding of
> nests and lower breeding success in the Baltic. The authors note that such
> flooding events are likely to further increase. Information regarding debris comes
> from personal correspondence with stakeholders.

2 - Range expansion of American mink, partly assisted by climate change, has led
to increased rates of predation at guillemot colonies

> Buchadas, A. R. C., and Hof, A. R. "Future breeding and foraging sites of a
> southern edge population of the locally endangered Black Guillemot
> Cepphus grylle." Bird Study 64.3 (2017): 306-316. Black guillemots are
> particularly vulnerable to predation by American mink, which is currently
> increasing in range and abundance in Europe. This range expansion has likely
> been assisted by climate change and therefore predation is likely to worsen
> across the species range. The study focusses on the Baltic but suggests
> anywhere the mink is expanding is likely to have similar issues in the future.

3 - Guillemot have shifted their laying date, most likely linked to an increase in sea
surface temperature and prey availability

> Greenwood, J. G. "Earlier laying by Black Guillemots Cepphus grylle in
> Northern Ireland in response to increasing sea-surface temperature." Bird
> Study 54.3 (2007): 378-379.

Atlantic Puffin

1 - Changes in puffins' prey availability during breeding season has led to decreased breeding success

Barrett, R. T. "Atlantic puffin Fratercula arctica and common guillemot Uria aalge chick diet and growth as indicators of fish stocks in the Barents Sea." Marine Ecology Progress Series 230 (2002): 275-287. The volume of puffin eggs in two Norwegian populations declined over a roughly 30-year period. Eggs were smaller in years when capelin and herring is less available, which is linked to climate change. This likely is also an indicator of population health and may be driving declines.

Burthe, S. J., et al. "Assessing the vulnerability of the marine bird community in the western North Sea to climate change and other anthropogenic impacts." Marine Ecology Progress Series 507 (2014): 277-295. Puffin productivity and survival has decreased around the Forth and Tay region as temperature has increased. Most likely linked to prey availability. Sea surface temperature has increased in the study region between 1980 and 2011.

Dunn, P. O., and Møller, A. P. (eds). "Effects of Climate Change on Birds" 2nd edition, Oxford, Oxford Academic (2019). Higher water temperature in the Norwegian sea has resulted in a shift in herring stock to the north, and a spatial mismatch between puffins and prey.

Durant, J. M., et al. "Regime shifts in the breeding of an Atlantic puffin population." Ecology Letters 7.5 (2004): 388-394. In the Norwegian sea, long term fluctuations in the winter NAO index have affected food availability and therefore puffin success and breeding timing. Study used a multi-decadal dataset from Hernyken, Northern Norway.

Fauchald, Per, et al. "The status and trends of seabirds breeding in Norway and Svalbard." NINA rapport 1151. Norsk institutt for naturforskning (2015): 1-84. Severe declines of puffins have occurred in most areas of Norway, cause is not known but changes in food abundance and timing are concluded to be the most probable cause.

Fayet, A. L., et al. "Local prey shortages drive foraging costs and breeding success in a declining seabird, the Atlantic puffin." Journal of Animal Ecology 90.5 (2021): 1152-1164. Puffin tracking data has shown that when they are forced to forage further (using data from Iceland, Norway and Wales) breeding success decreases. Changes in food availability – which the study suggests have caused puffins to forage further from the colony and expend more energy at foraging grounds – are closely related to temperature.

Frederiksen, M., et al. "Climate, copepods and seabirds in the boreal

Northeast Atlantic–current state and future outlook." Global Change Biology 19.2 (2013): 364-372. Declines in puffin breeding success on the Isle of May correlate with marine environmental suitability for copepods (a key prey for many fish and seabirds), which has decreased in recent years. The authors found a weaker, non-significant link in Norway.

Hansen, E. S., et al. "Centennial relationships between ocean temperature and Atlantic puffin production reveal shifting decennial trends." Global Change Biology 27 (2021): 3753-3764. Sea surface temperature is a strong predictor of puffin breeding success in a breeding population in Iceland, over both decennial and centennial timescales, most likely through sandeel abundance during the winters. Milder winters result in fewer sandeels in the following summer. Study uses a long term dataset (130 years) based on breeding success in Vestmannaeyjar, Iceland.

2 - Changes in puffins' prey availability during non-breeding season has led to increased mortality and population declines

Anon "Atlantic puffin Fratercula arctica" Scottish Wildlife Trust Report (2018) Available at: https://scottishwildlifetrust.org.uk/wp-content/uploads/2018/01/Puffin.docx. Colony declines on the east coast of Scotland are attributed to lack of prey during the breeding and non-breeding season. The report notes the availability of suitable prey during the non-breeding season is critical for long-term health of puffin populations.

Harris, M. P., et al. "Wintering areas of adult Atlantic puffins Fratercula arctica from a North Sea colony as revealed by geolocation technology." Marine Biology 157.4 (2010): 827-836. Between 1992 and 2008, Isle of May puffins suffered increasing over-winter mortality and shifted their wintering distribution from the North Sea to the Atlantic. The study suggests this might be linked to changes in temperature, plankton populations and fish populations in the North Sea.

3 - Changes in vegetation has led to fewer suitable puffin nest-sites

Burthe, S. J., et al. "Assessing the vulnerability of the marine bird community in the western North Sea to climate change and other anthropogenic impacts." Marine Ecology Progress Series 507 (2014): 277-295. Update on the below item confirming the impacts of tree mallow on puffins and the scale of the problem in Scotland.

Van Der Wal, R., et al. "Multiple anthropogenic changes cause biodiversity loss through plant invasion." Global Change Biology 14.6 (2008): 1428-1436. Expansion of tree mallow Lavatera arboreahas, in part due to climate change, substantially reduced suitable nesting habitat for Atlantic puffins at several colonies in the Forth and Tay region.

4 - Extreme storms during the non-breeding season have led to mass-mortality of puffins ('wrecks')

Clairbaux, M., et al. "North Atlantic winter cyclones starve seabirds." Current Biology 31.17 (2021): 3964-3971. Following heavy storm action, seabird mortality increases due to increased difficulty foraging (rather than increased energetic costs). The authors use a multi-species dataset (puffins, little auks, common murres, and thick-billed murres) over a wide area of the Atlantic basin. They conclude that seabirds around Iceland and the Barents Sea (along with several N. American sites) are particularly vulnerable. Climate change is likely to be a contributing factor to present and future storm mortality. **Mitchell, I., et al. "Impacts of climate change on seabirds, relevant to the coastal and marine environment around the UK." (2020): 382-399.** Winter storms can cause mass mortality (and have recently in 2013/14 storms), wrecks have been observed off the coast of France and the east coast of England. While individual storms cannot easily be attributed to climate change, most predictions are confident extreme Atlantic storms will become more frequent.

5 - Puffins have changed their wintering range

Harris, M. P., et al. "Wintering areas of adult Atlantic puffins Fratercula arctica from a North Sea colony as revealed by geolocation technology." Marine Biology 157.4 (2010): 827-836. Between 1992 and 2008, Isle of May puffins suffered increasing over-winter mortality and shifted their wintering distribution from the North Sea to the Atlantic. The study suggests this might be linked to changes in temperature, plankton populations and fish populations in the North Sea.

Common Murre

1 - High-wind events in the non-breeding season have led to mass mortality of murres in recent years

Louzao, M., et al. "Threshold responses in bird mortality driven by extreme wind events." Ecological Indicators 99 (2019): 183-192. High wind events in the winter have caused several mass mortality events ("wrecks") in NE Atlantic, though this study focusses on bodies washed up in Bay of Biscay. Study suggests high wind events are likely to become more common and result in more deaths in the future.

2 - Extreme storms during the non-breeding season have led to mass mortality of murres ('wrecks')

Clairbaux, M., et al. "North Atlantic winter cyclones starve seabirds."

Current Biology 31.17 (2021): 3964-3971. Following heavy storm action, seabird mortality increases due to increased difficulty foraging (rather than increased energetic costs). The authors use a multi-species dataset (puffins, little auks, common murres, and thick-billed murres) over a wide area of the Atlantic basin. They conclude that seabirds around Iceland and the Barents Sea (along with several N. American sites) are particularly vulnerable. Climate change is likely to be a contributing factor to present and future storm mortality.

3 - More frequent extreme storms during murres' breeding season has increased foraging difficulty and reduced food fed to chicks

Finney, S. K., Wanless, S., and Harris, M. P. "The effect of weather conditions on the feeding behaviour of a diving bird, the Common Guillemot Uria aalge." Journal of Avian Biology (1999): 23-30. Stormy weather affects the quantity and size of food that adults can provide to chicks, and increases in summer storm frequency may result in lowered foraging efficiency. Study was conducted on Isle of May. Extreme storms in the Atlantic are likely to become more frequent in the future and further disrupt the breeding season.

4 - Extreme storms during murres' breeding season have led to wide-spread nest destruction, nesting failure and a net reduction in annual population production

Newell, M., et al. "Effects of an extreme weather event on seabird breeding success at a North Sea colony." Marine Ecology Progress Series 532 (2015): 257-268. A single extreme summer storm on the Isle of May resulted in wide-spread nest destruction, nesting failure and a net reduction in annual population production. While individual storms cannot be easily be attributed to climate change, it is generally believed that severe storms are becoming more common

5 - Changes in murres' prey availability during the breeding season has led to decreased breeding success

Frederiksen, M., et al. "Climate, copepods and seabirds in the boreal Northeast Atlantic—current state and future outlook." Global Change Biology 19.2 (2013): 364-372. Breeding success on the Isle of May was strongly correlated to suitable climate for local copepods, and increases in temperature have lowered suitability for copepods and therefore breeding success in recent years. Projections also show that this drop in suitability will continue and worsen in the future. Interestingly, the authors did not find evidence of such a link in Rost, Norway.

Irons, D. B., et al. "Fluctuations in circumpolar seabird populations linked to climate oscillations." Global Change Biology 14.7 (2008): 1455-1463. The authors find, using a multi-decade dataset, that murre population size across the Arctic is strongly correlated to sea surface temperature. Rapid temperature shifts (either hotter or cooler) resulted in a decrease in population size. Note: this study does not explicitly investigate anthropogenic climate change, but does show a clear linkage between rapid climate change and population declines.

6 - Murres are more likely to skip breeding in warmer weather, and this behaviour is becoming more frequent. While this is a cause for concern, it is unclear what effect this will have on the population in the long-term

Reed, T. E., Harris, M. P., and Wanless, S. "Skipped breeding in common guillemots in a changing climate: restraint or constraint?" Frontiers in Ecology and Evolution 3 (2015): 1. Murres (aka guillemots) skip breeding more frequently in/following warm years, and sea temperatures have increased in the study area (North Sea) over recent decades. this may increase in the future. However, the long term consequences of this impact on the population are unclear. Study conducted on Isle of May, Scotland.

7 - Heatwaves have resulted in significant murre chick mortality. The frequency and severity of heatwaves is likely to increase

Ballstaedt, Elmar (personal correspondance); See also https://www. jordsand.eu/2018/08/14/bruterfolg-deutscher-seev%C3%B6gel-durch-wetterkapriolen-schlecht-wie-lange-nicht-mehr The heatwave of summer 2018 resulted in fewer breeding attempts and increased chick mortality in many species on Helgoland, most likely due to heat stress. Many seabirds on Helgoland are declining which could in part be due to climate change.

8 - Common murres have changed their phenology, potentially in response to climate change but the mechanism is unclear

Wanless, S., et al. "Later breeding in northern gannets in the eastern Atlantic." Marine Ecology Progress Series 370 (2008): 263-269. Over a period of roughly 30 years, the laying date of common murres in the North Sea (Isle of May and Farne Islands) became significantly later. The study did not identify a mechanism for delayed breeding in a warming North Sea, but did find a correlation between combined auk/kittiwake laying dates and the winter NAO index.

9 - A shift towards warmer, drier and calmer conditions has correlated with higher population abundance. Mechanism unknown, but likely mediated through prey availability and lower energetic costs.

Hemery, G., et al. "Detecting the impact of oceano-climatic changes on marine ecosystems using a multivariate index: the case of the Bay of

Biscay (North Atlantic-European Ocean)." Global Change Biology 14.1 (2008): 27-38. Abundance of common murres in the Bay of Biscay increased from 1974 to 2000. Annual abundance (at-sea counts) was positively correlated with a local multivariate climate index (combining 11 oceanic and atmospheric variables) and the large-scale winter NAO index. Murres appear to benefit from a trend towards warmer, drier years with calmer sea surface conditions.

Thick-billed Murre

1 - Changes in thick-billed murres' prey availability during the non-breeding season has led to increased mortality

Descamps, S., Strøm, H., and Steen, H.. "Decline of an arctic top predator: synchrony in colony size fluctuations, risk of extinction and the subpolar gyre." Oecologia 173.4 (2013): 1271-1282. Many colonies in Svalbard are declining, and if trends continue there is a risk of local extinction. Declines strongly correlate with marine change, though the exact mechanism is unknown. Authors suggest that deterioration of the feeding conditions in the winter affected bird survival, particularly juvenile survival, and that local variations in spring and summer conditions affected breeding propensity and breeding success of murres.

Sandvik, H., et al. "The effect of climate on adult survival in five species of North Atlantic seabirds." Journal of Animal Ecology 74.5 (2005): 817-831. Authors found that both the common murres (guillemots) and thick-billed murres (Brünnich's guillemots) were negatively affected by warmer temperatures causing alterations to their food webs, they note that these trends are likely to continue. The data spanned 14 years of observation at a colony on Hornøya, off Northern Norway in the western Barents Sea

2 - Changes in thick-billed murres' prey availability during the breeding season has led to decreased breeding success

Descamps, S., Strøm, H., and Steen, H.. "Decline of an arctic top predator: synchrony in colony size fluctuations, risk of extinction and the subpolar gyre." Oecologia 173.4 (2013): 1271-1282. Local variations in Svalbard spring and summer conditions affected breeding propensity and breeding success of murres. See 1) for more details.

Garðarsson, A., Guðmundsson, G. A., and Lilliendahl, K. "Svartfugl í íslenskum fuglabjörgum 2006–2008." Bliki 33 (2019): 35-46. Reviews the population trends in various seabird species in Iceland. Particularly highlights the drastic decline of thick-billed murres (Brünnich's guillemots) across Iceland. Sharp declines correspond to crash in key prey species and changes in marine ecosystems linked to rapid temperature change.

3 - Changes in thick-billed murres' prey availability during the breeding season has led to increased mortality

    Fluhr, J., et al. "Weakening of the subpolar gyre as a key driver of North Atlantic seabird demography: a case study with Brünnich's guillemots in Svalbard." Marine Ecology Progress Series 563 (2017): 1-11. An update and expansion on the previous paper, focussing on murres on Bear Island, Svalbard. Confirms strong correlation of adult annual survival and the strength of Atlantic subpolar gyre.

4 - Thick-billed murre populations are typically smaller and decline in areas with increasing sea temperatures. Mechanism unclear.

    Bonnet-Lebrun, A. S., et al. "Cold comfort: Arctic seabirds find refugia from climate change and potential competition in marginal ice zones and fjords." Ambio 51.2 (2022): 345-354. Thick-billed murres (Brünnich's guillemots) populations in Iceland have declined in correlation with rising sea surface temperatures. In addition, populations associated with higher sea temperatures have declined faster and tend to be smaller than those near refugia of cold water. The authors investigate the role of competition (with little significant effect), but link to various Icelandic studies which provide evidence for prey availability being the main reason behind the declines

    Irons, D. B., et al. "Fluctuations in circumpolar seabird populations linked to climate oscillations." Global Change Biology 14.7 (2008): 1455-1463. The authors find, using a multi-decade dataset, that murre population size across its range in the Arctic is strongly correlated to sea surface temperature. Rapid temperature shifts (either hotter or cooler) resulted in a decrease in population size, probably mediated through changes in underlying food-webs. Note: this study does not explicitly investigate anthropogenic climate change, but does show a clear linkage between rapid climate change and population declines.

5 - Extreme storms during the non-breeding season have led to mass mortality of murres ('wrecks')

    Clairbaux, M., et al. "North Atlantic winter cyclones starve seabirds." Current Biology 31.17 (2021): 3964-3971. Following heavy storm action, seabird mortality increases due to increased difficulty foraging (rather than increased energetic costs). The authors use a multi-species dataset (puffins, little auks, common murres, and thick-billed murres) over a wide area of the Atlantic basin. They conclude that seabirds around Iceland and the Barents Sea (along with several N. American sites) are particularly vulnerable. Climate change is likely to be a contributing factor to present and future storm mortality.

1.1.2 Change in European range size between present day and 2100:

Using a species distribution model (SDM) we correlated species occurrence during the breeding season with a number of terrestrial and marine environmental variables. Species range data came from the European Breeding Bird Atlas (EBBA2) database. Present-day and 2100 terrestrial data were downloaded from the WorldClim database. We used data from the MRI-ESM2 general circulation model (GCM), which is a high-performing model over Europe. Present-day and 2100 marine data were downloaded from the Bio-Oracle database which averages predictions of marine variables from several different atmospheric-oceanic general circulation models (AOGCMS; for full details see Assis et al., 2017). For the map presented in the summary we used representative concentration pathway (RCP) 4.5, which is an "intermediate" emissions scenario. All data were at 5-minute resolution.

For razorbill, little auk, black guillemot, Atlantic puffin, common murre, and thick-billed murre we included the following terrestrial variables: mean temperature of the warmest month, precipitation during breeding season, isolation of landmass, area of landmass, distance to sea.

For razorbill, little auk, black guillemot, Atlantic puffin, common murre, and thick-billed murre we included the following marine variables: sea surface temperature (during the winter), salinity, maximum chlorophyll concentration, bathymetry (depth and variance).

After running our model we generated a present-day map where every grid-cell is given a habitat suitability score between 0 and 1, where 1 is very suitable habitat and 0 is not at all suitable. We then compared this with a corresponding map built with 2100 data, and highlighted currently inhabited areas where 1) suitability drops sharply (i.e. by more than 0.1) and 2) suitability drops below a probability threshold set by the model. Conversely we also highlighted areas where suitability rose sharply and above a given threshold. While a drop in habitat suitability is likely to result in population declines, it is not a certainty, and it does not mean that a population will be extinct in 2100 or that a population is doomed to extinction. With conservation action and careful management, along with changes in human behaviour, such declines may be mitigated or in some cases prevented. For a full explanation of the model see the accompanying 'Methodology' folder in Appendix 2.

Underlying data were downloaded from:

Keller, V., et al. "European Breeding Bird Atlas 2: Distribution, Abundance and Change." European Bird Census Council & Lynx Edicions, Barcelona (2020). Source of range data

Fick, S. E., and Hijmans, R. J. "WorldClim 2: new 1-km spatial resolution climate surfaces for global land areas." International Journal of Climatology 37.12 (2017): 4302-4315. Source of present-day and 2100 terrestrial data.

Assis, J., et al. "Bio-ORACLE v2. 0: Extending marine data layers for bioclimatic modelling." Global Ecology and Biogeography 27.3 (2018): 277-284. Source of present-day and 2100 marine data

1.1.3 Changes in key prey species:

We first identified the key prey species for each species. This can be variable across a species' range, but if available evidence suggested at least one major population is highly dependent on a particular prey species, then typically this species would be included. Lists of prey species were compiled from published sources, then verified and expanded following consultation with conservation practitioners. Afterwards we compiled current and projected maps of prey ranges to assess where key prey species may become less common in the near future. If any of the key species are predicted to vanish or drastically reduce in abundance in the current foraging range a given species, we marked this on the summary map.

We used several sources to collate range information, but for preference we used data from COPERNICUS as they include projected abundance. For species where this was not available we used habitat suitability instead. In all cases we used RCP 4.5, which is an "intermediate" emissions scenario. For species in the COPERNICUS database we used the 0.6 maximum sustainable yield parameter, which assumes international co-operation to work towards fish-stock sustainability. Our assessment is therefore relatively conservative in terms of changes in prey species.

Razorbill key prey species: sandeel species (*Ammodytes marinus* and *Ammodytes tobianus*), herring (*Clupea harengus*), capelin (*Mallotus villosus*) and sprat (*Sprattus sprattus*). Prey species list was compiled from:

Barrett, R. T. "The diet, growth and survival of Razorbill Alca torda chicks in the southern Barents Sea." Ornis Norvegica 38 (2015): 25-31.

Fauchald, Per, et al. "The status and trends of seabirds breeding in Norway and Svalbard." NINA rapport 1151. Norsk institutt for naturforskning (2015): 1-84.

Little Auk key prey species: *Calanus glacialis*, *Calanus hyperboreus* and *Apherusa glacialis*. Note that since data regarding climate change and copepod range shifts are not readily available, a full prey loss assessment could not be carried out for this species. Prey species list was compiled from:

Amélineau, F., et al. "Arctic climate change and pollution impact little auk foraging and fitness across a decade." Scientific reports 9.1 (2019): 1-15.

Harding, A. M. A., et al. "Estimating prey capture rates of a planktivorous seabird, the little auk (Alle alle), using diet, diving behaviour, and energy consumption." Polar Biology 32.5 (2009): 785-796.

Black Guillemot key prey species: sandeel species (*Ammodytes marinus*), butterfish (*Pholis gunnellus*), eelpout (*Zoarces viviparus*) and sea scorpion (*Taurulus bubalis*). Prey species list was compiled from:

Barrett, R. T., and Anker-Nilssen, T. "Egg-laying, chick growth and food of Black Guillemots Cepphus grylle in North Norway." Fauna Norvegica, Series C 20.2 (1997): 69-79.

BirdLife International. "Species factsheet: Cepphus grylle." (2021) Downloaded from http://www.birdlife.org on 01/06/2021.

Ewins, P. J. "The diet of black guillemots Cepphus grylle in Shetland." Ecography 13.2 (1990): 90-97.

Fauchald, Per, et al. "The status and trends of seabirds breeding in Norway and Svalbard." NINA rapport 1151. Norsk institutt for naturforskning (2015): 1-84.

Hario, M. "Chick growth and nest departure in Baltic Black Guillemots Cepphus grylle." Ornis Fennica 78.3 (2001): 97-108.

Hof, A. R., Crombag, J. A. H. M. , and Allen, A. M. "The ecology of Black Guillemot Cepphus grylle grylle chicks in the Baltic Sea region: insights into their diet, survival, nest predation and moment of fledging." Bird Study 65.3 (2018): 357-364.

Atlantic Puffin key prey species: sandeel species (*Ammodytes marinus* and *Ammodytes tobianus*), herring (*Clupea harengus*), capelin (*Mallotus villosus*) and sprat (*Sprattus sprattus*). Prey species list was compiled from:

BirdLife International. "Species factsheet: Fratercula arctica." (2021) Downloaded from http://www.birdlife.org on 01/06/2021.

Fayet, A. L., et al. "Local prey shortages drive foraging costs and breeding success in a declining seabird, the Atlantic puffin." Journal of Animal Ecology 90.5 (2021): 1152-1164.

Common Murre key prey species: sprat (*Sprattus sprattus*), sandeels species (*Ammodytes marinus* and *Ammodytes tobianus*), capelin (*Mallotus villosus*), herring

(*Clupea harengus*), Atlantic cod (*Gadus morhua*), saithe (*Pollachius virens*) and haddock (*Melanogrammus aeglefinus*). Prey species list was compiled from:

Ainley, D. G., Nettleship, D. N., and Storey, A. E. "Common Murre (Uria aalge), version 2.0." In Birds of the World (S. M. Billerman, P. G. Rodewald, and B. K. Keeney, Editors). Cornell Lab of Ornithology, Ithaca, NY, USA (2021).

BirdLife International. "Species factsheet: Uria aalge." (2021) Downloaded from http://www.birdlife.org on 01/06/2021.

Fauchald, Per, et al. "The status and trends of seabirds breeding in Norway and Svalbard." NINA rapport 1151. Norsk institutt for naturforskning (2015): 1-84.

Thick-billed Murre key prey species: sandeel species (*Ammodytes marinus* and *Ammodytes tobianus*), herring (*Clupea harengus*), capelin (*Mallotus villosus*), Atlantic cod (*Gadus morhua*) and polar cod (*Boreogadus saida*). Prey species list was compiled from:

Fauchald, Per, et al. "The status and trends of seabirds breeding in Norway and Svalbard." NINA rapport 1151. Norsk institutt for naturforskning (2015): 1-84.

Prey range information for all species were compiled from:

Kesner-Reyes, K., et al. "AquaMaps: Predicted range maps for aquatic species." In FishBase: R. Froese & D. Pauly (Eds.) (2019). Available at: https://www.aquamaps.org

Sailley, S., et al. "Fish abundance and catch data for the Northwest European Shelf and Mediterranean Sea from 2006 to 2098 derived from climate projections". Copernicus Climate Change Service (C3S) Climate Data Store (CDS) (2021). https://doi.org/10.24381/cds.39c97304.

1.1.4 Climate change impacts outside of Europe
Little Auk

Loss of sea ice and new prey items due to climate change has led to increased little auk breeding success in Greenland

Amélineau, F., et al. "Arctic climate change and pollution impact little auk foraging and fitness across a decade." Scientific reports 9.1 (2019): 1-15.

Atlantic Puffin

Some colonies in North America have changed their laying phenology, presumably in response to temperature and/or prey availability. Some recent observations have reported this has also occurred in Europe.

Major, H. L., et al. "Contrasting phenological and demographic responses of Atlantic Puffin (Fratercula arctica) and Razorbill (Alca torda) to climate change in the Gulf of Maine." Elem Sci Anth 9.1 (2021): 00033.

Erpur Hansen (Personal Correspondence)

Thick-billed Murre

Thick-billed murres are known to be impacted by climate change outside of Europe. Impacts include increased predation by polar bears, increased parasitism by mosquitoes (leading to breeding failure), and increased mortality caused by algal blooms. Changes in the marine ecosystem in the canadian high Arctic, driven by climate change, has resulted in higher concentrations of mercury bioaccumulated in thick-billed murres. No long-term impact on population health has been observed so far.

Braune, B. M., et al. "Changes in food web structure alter trends of mercury uptake at two seabird colonies in the Canadian Arctic." Environmental science & technology 48.22 (2014): 13246-13252.

Gaston, A. J., and Elliott, K. H. "Effects of climate-induced changes in parasitism, predation and predator-predator interactions on reproduction and survival of an Arctic marine bird." Arctic (2013): 43-51.

Kuletz, K. et al. "Chapter 3.5: Seabirds" in "State of the Arctic Marine Biodiversity Report". Conservation of Arctic Flora and Fauna International Secretariat (2017): 129-147.

1.2 Sensitivity (references)

We used a list of candidate traits based on that in Foden & Young (2016) that indicate high sensitivity and identified which, if any, auks possessed. In brief, we consulted published literature as well as expert knowledge and online databases such as Birdlife (http://datazone.birdlife.org/) and Birds of the World (https://birdsoftheworld.org), to assess whether auks have either 1) Specialised habitat and/or microhabitat requirement 2) Environmental tolerances or thresholds (at any life stage) that are likely to be exceeded due to climate change 3) Dependence on environmental triggers that are likely to be disrupted by climate change, 4) Dependence on interspecific interactions that are likely to be disrupted by climate change or 5) High rarity.

For more detail and a full list of traits see:

Foden, W. B. and Young, B. E. (eds.). "IUCN SSC Guidelines for Assessing Species' Vulnerability to Climate Change. Version 1.0." Occasional Paper of the IUCN Species Survival Commission No. 59 (2016). Cambridge, UK and Gland, Switzerland: IUCN Species Survival Commission. x+114pp.

1.3 Adaptive capacity (references)

We used a list of candidate traits based on that in Foden & Young (2016) that indicate adaptive capacity and identified which, if any, auks possessed. In brief, we consulted published literature as well as expert knowledge and online databases such as Birdlife (http://datazone.birdlife.org/) and Birds of the World (https://birdsoftheworld.org), to assess whether auks have either: 1) High phenotypic plasticity. 2) High dispersal ability or 3) High evolvability.

For more detail and a full list of traits see:

Foden, W. B. and Young, B. E. (eds.). "IUCN SSC Guidelines for Assessing Species' Vulnerability to Climate Change. Version 1.0." Occasional Paper of the IUCN Species Survival Commission No. 59 (2016). Cambridge, UK and Gland, Switzerland: IUCN Species Survival Commission. x+114pp.

© Seppo Häkkinen

Appendix 1: Ducks and Phalaropes
Sources and references for vulnerability assessment

1.1 Evidence for exposure (references)

1.1.1 Current impacts attributed to climate change:

Long-tailed Duck

1 - Wintering populations in Europe have declined due to climate change-driven changes in predation in breeding areas outside of Europe.

Hario, M., Rintala, J., and Nordenswan, G. "Dynamics of wintering long-tailed ducks in the Baltic Sea–the connection with lemming cycles, oil disasters, and hunting." Suomen Riista 55 (2009): 83-96. Wintering populations in the Baltic have rapidly declined, in part because of the effects of climate change on key breeding sites outside of Europe. Due to changes in lemming availability, predators such as Arctic foxes have swapped to preying on duck eggs and young as alternative prey which, in turn, has resulted in decreased breeding success of long-tailed duck Clangula hyemalis on the Taimyr Peninsula.

Hearn, R. D., Harrison, A. L., and Cranswick, P. A. "International single species action plan for the conservation of the long-tailed duck Clangula hyemalis, 2016–2025." AEWA Technical Series Species Action Plan: 57 (2015). Expands and updates the work above, the authors believe changes in predation have affected populations across western Siberia and northern Europe

2 - Range expansion of red foxes following milder winters has led to predation of ducks much further north than previously, and may be threatening the viability of northern populations.

Hearn, R. D., Harrison, A. L., and Cranswick, P. A. "International single species action plan for the conservation of the long-tailed duck Clangula hyemalis, 2016–2025." AEWA Technical Series Species Action Plan: 57 (2015). Range expansion of predators such as red fox (Vulpes vulpes) may be influencing predator-prey relationships in the Arctic breeding grounds, and appears to be threatening the viability of the small Finnish Lapland breeding population of long-tailed ducks. Additional information on the risk and spread of red foxes due to climate change was provided by stakeholders.

3 - Competition with non-native gobies has caused long-tailed ducks to switch prey, though there has been no observed change in mortality or condition. Goby invasion

may have been assisted by climate change, though currently this is speculative.

Behrens, J. W., et al. "Seasonal depth distribution and thermal experience of the non-indigenous round goby Neogobius melanostomus in the Baltic Sea: implications to key trophic relations." Biological Invasions 24.2 (2022): 527-541. This study does not make a link directly to seabirds, but is provided as supplementary information making the link between gobies and climate change clearer. Gobies strongly prefer warm water, which explain why they have now colonised the Baltic as climate change has resulted in warmer winters. It also suggests further climate change will assist further spread.

Skabeikis, A., et al. "Effect of round goby (Neogobius melanostomus) invasion on blue mussel (Mytilus edulis trossulus) population and winter diet of the long-tailed duck (Clangula hyemalis)." Biological Invasions 21.3 (2019): 911-923. The benthic round goby has recently colonised the Baltic, which may have been facilitated by climate change (Ramunas Žydelis, personal communication), as gobies strongly prefer warmer water (see Behrens et al. 2022 below). Competition with gobies has caused long-tailed ducks to switch prey, and there has been no observed change in mortality or condition. However, further climate change could promote goby expansion and further competition.

Harlequin Duck

1 - Population has redistributed, with some populations growing and others shrinking, most likely due to shifts in prey species caused by climate change

Gardarsson, A. "Harlequin Ducks in Iceland." Waterbirds 31.sp2 (2008): 8-14.

Gardarsson, A., and Einarsson, Á. "Relationships among food, reproductive success and density of harlequin ducks on the River Laxá at Myvatn, Iceland (1975-2002)." Waterbirds 31.sp2 (2008): 84-91. Using a multidecade dataset the authors conclude that southern populations in Iceland have decreased 1961-2001 while northern populations have increased. This is likely due to changes in blackfly abundance, which in turn is at least partly due to warmer springs and summers

Velvet Scoter

1 - Scoters are starting their autumn migration significantly later in response to changing climate.

Lehikoinen, A., and Jaatinen, K. "Delayed autumn migration in northern European waterfowl." Journal of Ornithology 153.2 (2012): 563-570. Scoter phenology changed between 1979 and 2009, consistent with expectations under a warming local climate. Autumn migration is occurring later, birds are arriving later. Study carried out using long term data from Hanko Observatory, southern Finland

2 - Wintering populations have redistributed, most likely due to lack of prey caused at least partly by climate change.

Tolon, V., et al. "Etat des populations de macreuses en Europe, en France et en Basse-Normandie et analyse des principaux facteurs de distribution". Report for Maison de l'Estuaire (2013). Wintering populations of scoters off the coast of France are declining and in some cases have disappeared. Cause is uncertain, but probably they have redistributed rather than died, and have shifted in response to reduced prey availability. There are several underlying causes, but climate change is likely to be a contributing factor.

Common Scoter

1 - Wintering populations have redistributed, most likely due to lack of prey caused at least partly by climate change.

Tolon, V., et al. "Etat des populations de macreuses en Europe, en France et en Basse-Normandie et analyse des principaux facteurs de distribution". Report for Maison de l'Estuaire (2013). Wintering populations of scoters off the coast of France are declining and in some cases have disappeared. Cause is uncertain, but probably they have redistributed rather than died, and have shifted in response to reduced prey availability. There are several underlying causes, but climate change is likely to be a contributing factor.

Red-necked Phalarope

1 - Red-necked phalaropes have shifted north in Finland, the most southerly populations are declining while northerly populations are increasing. This shift is in correlation with climate change, but the underlying mechanism is not certain

Virkkala, R., et al. "Matching trends between recent distributional changes of northern-boreal birds and species-climate model predictions." Biological Conservation 172 (2014): 124-127. Red-necked phalaropes have shifted north in Finland, the central density of the population has shifted significantly northwards. This shift is in correlation with climate change, but the underlying mechanism is not certain.

Steller's Eider

1 - Many Steller's eiders in the Baltic have changed wintering area to the White Sea, most likely due to decreases in sea ice. This may also be associated with an overall population decline, but this is uncertain

Aarvak, T., et al. "The European wintering population of Steller's Eider Polysticta stelleri reassessed." Bird Conservation International 23.3 (2013):

337-343. The number of Steller's eiders wintering in the Baltic has dropped sharply. This is likely due to both a decrease in population size and a redistribution of wintering area to the White sea, most likely due to decreases in sea ice and a greater area of open water.

> **Żydelis, R., et al. "Recent changes in the status of Steller's Eider Polysticta stelleri wintering in Europe: a decline or redistribution?." Bird Conservation International 16.3 (2006): 217-236.** This paper precedes the one above and concludes that although there was a redistribution towards the Kola Peninsula, the population may be declining as a whole.

Common Eider

1 - Milder winter and summer weather have resulted in better average adult condition, and therefore better breeding success. In some areas this has resulted in local populations increases.

> **D'Alba, L., Monaghan, P., and Nager, R. G. "Advances in laying date and increasing population size suggest positive responses to climate change in common eiders Somateria mollissima in Iceland." Ibis 152.1 (2010): 19-28.** Using a 30 year dataset in Iceland, the authors believe climate change is a major driver behind the population increase. Milder summers mean more nests, because fewer females skip breeding (as they are in higher condition)

2 - Eiders have shifted their phenology in response to milder winters and lay earlier.

> **D'Alba, L., Monaghan, P., and Nager, R. G. "Advances in laying date and increasing population size suggest positive responses to climate change in common eiders Somateria mollissima in Iceland." Ibis 152.1 (2010): 19-28.** In Iceland, eiders laid earlier following warmer winters. The exact reason is uncertain but could be because adults are in better condition following winter, or because key prey species (especially mussels) are available earlier in milder winters.

3 - Due to a lack of sea ice driven by climate change, polar bears are becoming more numerous around bird colonies during the summer and are more heavily predating on eider populations

> **Prop, J., et al. "Climate change and the increasing impact of polar bears on bird populations." Frontiers in Ecology and Evolution 3 (2015): 33.** Study conducted on Svalbard, polar bears appear to be swapping prey species from seals due to a lack of sea ice. Several bird species are increasingly predated, prominently eiders.

4 - Earlier melt of sea ice in spring has resulted in a decrease in predation by Arctic

foxes, as they cannot access breeding colonies without the presence of sea ice

> **Hanssen, S. A., et al. "A natural antipredation experiment: predator control and reduced sea ice increases colony size in a long-lived duck." Ecology and Evolution 3.10 (2013): 3554-3564.** Across two eider colonies in Spitsbergen, population density is greater in years with less sea ice in April. The study suggests this provides earlier/longer access to food and reduces predation by Arctic foxes. The study used a 30-year dataset, and notes that April sea ice cover has declined over this period.

5 - Earlier melt of sea ice in spring has resulted in an increase in eider population density, as eiders have earlier and longer access to high-quality prey.

> **Hanssen, S. A., et al. "A natural antipredation experiment: predator control and reduced sea ice increases colony size in a long-lived duck." Ecology and Evolution 3.10 (2013): 3554-3564.** Across two eider colonies in Spitsbergen, population density is greater in years with less sea ice in April. The study suggests this provides earlier/longer access to food and reduces predation by Arctic foxes. The study used a 30-year dataset, and notes that April sea ice cover has declined over this period.

1.1.2 Change in European range size between present day and 2100:

Using a species distribution model (SDM) we correlated species occurrence during the breeding season with a number of terrestrial and marine environmental variables. Species range data came from the European Breeding Bird Atlas (EBBA2) database. Present-day and 2100 terrestrial data were downloaded from the WorldClim database. We used data from the MRI-ESM2 general circulation model (GCM), which is a high-performing model over Europe. Present-day and 2100 marine data were downloaded from the Bio-Oracle database which averages predictions of marine variables from several different atmospheric-oceanic general circulation models (AOGCMS; for full details see Assis et al., 2017). For the map presented in the summary we used representative concentration pathway (RCP) 4.5, which is an "intermediate" emissions scenario. All data were at 5-minute resolution.

For long-tailed duck, harlequin duck, velvet scoter, common scoter, red-breasted merganser, red phalarope, and red-necked phalarope we included the following terrestrial variables: Mean temperature of the warmest month, precipitation during breeding season, distance to sea

For Steller's eider, common eider, and king eider we included the following terrestrial variables: mean temperature of the warmest month, precipitation during breeding season, isolation of landmass, area of landmass, distance to sea.

For Steller's eider, common eider, and king eider we included the following marine variables: sea surface temperature (during the winter), salinity, maximum chlorophyll concentration, bathymetry (depth and variance).

Several other variables may strongly influence the distribution of ducks and phalaropes and it is not possible to include all possible variables in a given model. However the following variables have previously been found to be important to predicting the distribution of ducks and phalaropes in Europe: freshwater depth, freshwater ph, freshwater chlorophyll concentration, seabed substrate (sediment granulometry). For local assessments of climate change, we recommend these variables are strongly considered. We hope to incorporate these variables into future versions of this resource.

After running our model we generated a present-day map where every grid-cell is given a habitat suitability score between 0 and 1, where 1 is very suitable habitat and 0 is not at all suitable. We then compared this with a corresponding map built with 2100 data, and highlighted currently inhabited areas where 1) suitability drops sharply (i.e. by more than 0.1) and 2) suitability drops below a probability threshold set by the model. Conversely we also highlighted areas where suitability rose sharply and above a given threshold. While a drop in habitat suitability is likely to result in population declines, it is not a certainty, and it does not mean that a population will be extinct in 2100 or that a population is doomed to extinction. With conservation action and careful management, along with changes in human behaviour, such declines may be mitigated or in some cases prevented. For a full explanation of the model see the accompanying 'Methodology' folder in Appendix 2.

Underlying data were downloaded from:

> Keller, V., et al. "European Breeding Bird Atlas 2: Distribution, Abundance and Change." European Bird Census Council & Lynx Edicions, Barcelona (2020). Source of range data

> Fick, S. E., and Hijmans, R. J. "WorldClim 2: new 1-km spatial resolution climate surfaces for global land areas." International Journal of Climatology 37.12 (2017): 4302-4315. Source of present-day and 2100 terrestrial data.

> Assis, J., et al. "Bio-ORACLE v2. 0: Extending marine data layers for bioclimatic modelling." Global Ecology and Biogeography 27.3 (2018): 277-284. Source of present-day and 2100 marine data

1.1.3 Changes in key prey species:

We first identified the key prey species for each species. This can be variable across a species' range, but if available evidence suggested at least one major population is highly dependent on a particular prey species, then typically this species would be included. Lists of prey species were compiled from published sources, then verified and expanded following consultation with conservation practitioners. Afterwards we compiled current and projected maps of prey ranges to assess where key prey species may become less common in the near future. If any of the key species are predicted to vanish or drastically reduce in abundance in the current foraging range a given species, we marked this on the summary map.

We used several sources to collate range information, but for preference we used data from COPERNICUS as they include projected abundance. For species where this was not available we used habitat suitability instead. In all cases we used RCP 4.5, which is an "intermediate" emissions scenario. For species in the COPERNICUS database we used the 0.6 maximum sustainable yield parameter, which assumes international co-operation to work towards fish-stock sustainability. Our assessment is therefore relatively conservative in terms of changes in prey species.

Long-tailed Duck key prey species: This species relies on predominantly aquatic invertebrates on breeding grounds, and a variety of invertebrates, notably Mytilus species, and fish in winter. No key species could be identified so currently there is no key prey assessment for this species

Harlequin Duck key prey species: In summer, this species preys mainly on various midges, blackfly and caddis flies. In winter, no key species could be identified as diet is extremely varied. Currently this species does not have a key prey assessment

Velvet Scoter key prey species: *Mya arenaria, Cerastoderma glaucum, Saduria entomon, Euspira nitida, Macoma baltica, Cerastoderma lamarcki and Spisula subtruncata*. Prey species list was compiled from:

> Morkune, R., et al. "Triple stable isotope analysis to estimate the diet of the Velvet Scoter (Melanitta fusca) in the Baltic Sea." PeerJ 6 (2018): e5128.

> Durinck, J, et al. "Diet of the common scoter Melanitta nigra and velvet scoter Melanitta fusca wintering in the North Sea." Ornis Fennica 70.4 (1993): 215-218.

> Stempniewicz, L. "The food intake of two Scoters Melanitta fusca and M. nigra wintering in the Gulf of Gdańsk, Polish Baltic coast." Vår Fågelv., Suppl 11 (1986): 211-214..

Common Scoter key prey species: *Spisula subtruncata, Mya truncata, Macoma* *balthica, Mytilis edulis and Donax vittatus.* This species also preys on insects, especially chironomid larvae and cladocerans. Presently these are not included in the key prey assessment, due to data limitations.. Prey species list was compiled from:

Carboneras, C. and Kirwan, G. M. "Common Scoter (Melanitta nigra), version 1.0." In Birds of the World (J. del Hoyo, A. Elliott, J. Sargatal, D. A. Christie, and E. de Juana, Editors). Cornell Lab of Ornithology, Ithaca, NY, USA (2020).

Durinck, J., et al. "Diet of the common scoter Melanitta nigra and velvet scoter Melanitta fusca wintering in the North Sea." Ornis Fennica 70.4 (1993): 215-218.

Hartley, C. "Status and distribution of Common Scoters on the Solway Firth." British Birds 100.5 (2007): 280.

Stempniewicz, L. "The food intake of two Scoters Melanitta fusca and M. nigra wintering in the Gulf of Gdansk, Polish Baltic coast." Vår Fågelv., Suppl 11 (1986): 211-214.

Red-breasted Merganser key prey species: stickleback (*Gasterosteus aculeatus*). This species consumes a wide variety of fish species, both marine and freshwater. Freshwater species, especially *Salmo salar,* are likely very important but freshwater species are currently not included in the key prey assessment. While it preys on other marine species, no other key species could be identified. Prey species list was compiled from:

Bengtson, S.-A. "Food and feeding of diving ducks breeding at Lake Myvatn, Iceland." Ornis Fennica Vol. 48 (1971): 77-92.

Craik, S., Pearce, J., and Titman, R. D. "Red-breasted Merganser (Mergus serrator), version 1.0." In Birds of the World (S. M. Billerman, Editor). Cornell Lab of Ornithology, Ithaca, NY, USA (2020).

Feltham, M. "The diet of red-breasted mergansers (Mergus serrator) during the smolt run in NE Scotland: the importance of salmon (Salmo salar) smolts and parr." Journal of Zoology 222.2 (1990): 285-292.

Red Phalarope key prey species: This species has a broad diet of invertebrates, that varies across populations. Key species groups include marine copepods and amphipods, as well as adult and larval midges, gnats and craneflies. Currently there is no key prey assessment, due to lack of data.

Red-necked Phalarope key prey species: During the breeding species this species feed primarily on midges (adults and larvae), along with many other insect species. At sea, it feeds mostly on copepods and krill species. Currently there is no key prey assessment for this species

Steller's Eider key prey species: *Margarites helicinus, Skeneopsis planorbis, Mytilus edulis, Turtonia minuta, Gammarus oceanicus, Ampithoe rubricata, Idotea emarginata and Idotea granulosa*. This species also preys on various midge and cranefly larvae, especially during the breeding season. These terrestrial species are not included in the key prey assessment. Prey species list was compiled from

> Bustnes, Jan O., et al. "The diet of Steller's Eiders wintering in Varangerfjord, northern Norway." The Wilson Journal of Ornithology 112.1 (2000): 8-13.

> Fredrickson, L. H. "Steller's Eider (Polysticta stelleri), version 1.0." In Birds of the World (S. M. Billerman, Editor). Cornell Lab of Ornithology, Ithaca, NY, USA (2020).

> Nygård, T., Frantzen, B., and Švažas, Saulius. "Steller's Eiders Polysticta stelleri wintering in Europe: numbers, distribution and origin." Wildfowl 46.46 (1995): 140-156.

Common Eider key prey species: *Mytilus edulis, Modiolus modiolus, Tonicella marmorea, Buccinum undatum, Hyas araneus and Lacuna vincta*. Prey species list was compiled from:

> Bustnes, J. O., and Erikstad, K. E. "The diets of sympatric wintering populations of Common Eider Somateria mollissima and King Eider S. spectabilis in Northern Norway." Ornis Fennica 65.4 (1988): 163-168.

> Goudie, R. I., Robertson, G. J., and Reed, A. "Common Eider (Somateria mollissima), version 1.0". In Birds of the World (S. M. Billerman, Editor). Cornell Lab of Ornithology, Ithaca, NY, USA (2020).

> Kristjánsson, T. Ö., Jónsson, J. E., and Svavarsson, J. "Spring diet of common eiders (Somateria mollissima) in Breiðafjörður, West Iceland, indicates non-bivalve preferences." Polar Biology 36.1 (2013): 51-59.

King Eider key prey species: *Ophiopholis aculeata, Strongylocentrotus droebachiensis, Asterias rubens, Boreotrophon clathratus, Musculus discors, Modiolaria modiolus, Chlamys islandica, Mya truncata, Mytilus edulis, Ciliatocardium ciliatum and Hiatella arctica*. Prey species list was compiled from:

> Powell, A. N. and Suydam, R. S. "King Eider (Somateria spectabilis), version 1.0." In Birds of the World (S. M. Billerman, Editor). Cornell Lab of

Ornithology, Ithaca, NY, USA (2020).

Frimer, O. "Diet of moulting king eiders Somateria spectabilis at Disko Island, West Greenland." Ornis Fennica 74 (1997): 187-194.

Merkel, F. R., et al. "The diet of king eiders wintering in Nuuk, Southwest Greenland, with reference to sympatric wintering common eiders." Polar Biology 30.12 (2007): 1593-1597.

Prey range information for all species were compiled from:

Kesner-Reyes, K., et al. "AquaMaps: Predicted range maps for aquatic species." In FishBase: R. Froese & D. Pauly (Eds.) (2019). Available at: https://www.aquamaps.org

Sailley, S., et al. "Fish abundance and catch data for the Northwest European Shelf and Mediterranean Sea from 2006 to 2098 derived from climate projections". Copernicus Climate Change Service (C3S) Climate Data Store (CDS) (2021). https://doi.org/10.24381/cds.39c97304.

Velvet Scoter

Climate change has contributed to declines of scoter populations in North America. Earlier spring snow melt has likely led to a trophic mismatch and lower breeding success in scoters.

Drever, M. C., et al. "Population vulnerability to climate change linked to timing of breeding in boreal ducks." Global Change Biology 18.2 (2012): 480-492.

Common Scoter

Climate change has contributed to declines of scoter populations in North America. Earlier spring snow melt has likely led to a trophic mismatch and lower breeding success in scoters.

Drever, M. C., et al. "Population vulnerability to climate change linked to timing of breeding in boreal ducks." Global Change Biology 18.2 (2012): 480-492.

Red Phalarope

In Alaska red phalaropes now lay smaller eggs on average, presumably due to lower condition. This is likely due to delayed snow melt due to higher precipitation, despite the general warming trend. Across California red phalaropes have declined across their wintering areas. This is likely due to changes in ocean currents and declines in prey abundance. Populations around Alaska have declined in some areas, or possibly redistributed, due to changes in sea ice and in key copepod prey species

Gall, A. E., et al. "Ecological shift from piscivorous to planktivorous seabirds in the Chukchi Sea, 1975–2012." Polar Biology 40.1 (2017): 61-78.

Martin, J.-L., et al. "Late snowmelt can result in smaller eggs in Arctic shorebirds." Polar Biology 41.11 (2018): 2289-2295.

Sydeman, W. J., et al. "Climate–ecosystem change off southern California: time-dependent seabird predator–prey numerical responses." Deep Sea Research Part II: Topical Studies in Oceanography 112 (2015): 158-170.

Red-necked Phalarope

A study in Alaska found that phalaropes have changed their laying date in response to changes in snow melt. Phalaropes have responded to changes in oceanic patterns in the Indian ocean and changed their foraging areas and patterns in response.

Liebezeit, J. R., et al. "Phenological advancement in arctic bird species: relative importance of snow melt and ecological factors." Polar Biology 37.9 (2014): 1309-1320.

Nussbaumer, R., et al. "Investigating the influence of the extreme Indian Ocean Dipole on the 2020 influx of Red-necked Phalaropes Phalaropus lobatus in Kenya." Ostrich (2021): 1-9.

Common Eider

Known to be affected by climate change in other parts of their range. They suffer increased predation from Arctic foxes due to prey switching following a collapse in lemming breeding cycles in northern Canada. In addition Canadian populations have suffered due to changes in weather in the breeding season, especially increased rain, either directly through exposure or indirectly through changes in predation.

Iles, D. T., et al. "Predators, alternative prey and climate influence annual breeding success of a long-lived sea duck." Journal of Animal Ecology 82.3 (2013): 683-693.

King Eider

Increase in ice break-up, and increased variability of break-up, caused by climate change has resulted in significant damage to benthic prey and has caused local shifts in prey availability. Currently this has only a small impact on king eiders, but impacts could become significant in the future.

Lovvorn, J. R., et al. "Limits to benthic feeding by eiders in a vital Arctic migration corridor due to localized prey and changing sea ice." Progress in Oceanography 136 (2015): 162-174.

1.2 Sensitivity (references)

We used a list of candidate traits based on that in Foden & Young (2016) that indicate high sensitivity and identified which, if any, sea ducks and phalaropes possessed. In brief, we consulted published literature as well as expert knowledge and online databases such as Birdlife (http://datazone.birdlife.org/) and Birds of the World (https://birdsoftheworld.org), to assess whether sea ducks and phalaropes have either 1) Specialised habitat and/or microhabitat requirement 2) Environmental tolerances or thresholds (at any life stage) that are likely to be exceeded due to climate change 3) Dependence on environmental triggers that are likely to be disrupted by climate change, 4) Dependence on interspecific interactions that are likely to be disrupted by climate change or 5) High rarity. For more detail and a full list of traits see:

Foden, W. B. and Young, B. E. (eds.). "IUCN SSC Guidelines for Assessing Species' Vulnerability to Climate Change. Version 1.0." Occasional Paper of the IUCN Species Survival Commission No. 59 (2016). Cambridge, UK and Gland, Switzerland: IUCN Species Survival Commission. x+114pp.

1.3 Adaptive capacity (references)

We used a list of candidate traits based on that in Foden & Young (2016) that indicate adaptive capacity and identified which, if any, sea ducks and phalaropes possessed. In brief, we consulted published literature as well as expert knowledge and online databases such as Birdlife (http://datazone.birdlife.org/) and Birds of the World (https://birdsoftheworld.org), to assess whether sea ducks and phalaropes have either: 1) High phenotypic plasticity. 2) High dispersal ability or 3) High evolvability.

For more detail and a full list of traits see:

Foden, W. B. and Young, B. E. (eds.). "IUCN SSC Guidelines for Assessing Species' Vulnerability to Climate Change. Version 1.0." Occasional Paper of the IUCN Species Survival Commission No. 59 (2016). Cambridge, UK and Gland, Switzerland: IUCN Species Survival Commission. x+114pp.

© Silviu Petrovan

Appendix 1: Gannets and Cormorants

Sources and references for vulnerability assessment

1.1 Evidence for exposure (references)

1.1.1 Current impacts attributed to climate change:

Northern Gannet

1 - Gannets are undertaking longer foraging trips, most likely in response to prey shortages due to climate change. Although this likely increases the energetic costs of foraging, there have so far been no observed impacts on breeding success or mortality.

> Davies, R. D., et al. "Density-dependent foraging and colony growth in a pelagic seabird species under varying environmental conditions." Marine Ecology Progress Series 485 (2013): 287-294. Gannets are undertaking longer and longer foraging trips Celtic and Irish Seas. This is presumably due to food shortages closer to the colony, and climate change is likely one driver of this. Colonies are still growing and expanding, but at a slower rate than colonies elsewhere in the UK. This likely means these colonies will be more sensitive in future to climate change, but at the moment this is not classed as a negative impact.

2 - Gannets have established new colonies as key prey species have shifted further north.

> Barrett, R. T., Strøm, H., and Melnikov, M. "On the polar edge: the status of the northern gannet (Morus bassanus) in the Barents Sea in 2015-16." Polar Research 36.1 (2017): 1390384. Since 2011 gannets have established colonies in the Barents Sea (first on Bear Island), thought to be associated with a warming of the Barents Sea and the northward spread of common prey species.

European Shag

1 - Shags have advanced their laying date, most likely due to changes in marine and terrestrial temperatures and subsequently in prey availability

> Álvarez, D., and Pajuelo, M. A. F. "Southern populations of European shag Phalacrocorax a. aristotelis advance their laying date in response to local weather conditions but not to large-scale climate." Ardeola 58.2 (2011):

239-250. Laying dates for shags in northern Spain have changed drastically, advancing by almost 40 days in only 10 years. This is in correlation with an increase in local land and ocean temperatures, which is also the most likely reason behind this change in phenology.

2 - The diet composition of shags has changed a great deal, likely in response to climate change driven changes in the marine ecosystem

Howells, R. J., et al. "Pronounced long-term trends in year-round diet composition of the European shag Phalacrocorax aristotelis." **Marine Biology 165.12 (2018): 1-15**. Diet composition has changed, most likely in response to climate change. No known impact on population. Study on the Isle of May

Howells, R. J., et al. "From days to decades: short-and long-term variation in environmental conditions affect offspring diet composition of a marine top predator." **Marine Ecology Progress Series 583 (2017): 227-242**. Diet composition has changed, in correlation to climate change and change in sandeel abundance. Study on the Isle of May.

3 - Extreme storms during the shag breeding season have led to wide-spread nest destruction, nesting failure and a net reduction in annual population production

Newell, M., et al. "Effects of an extreme weather event on seabird breeding success at a North Sea colony." **Marine Ecology Progress Series 532 (2015): 257-268** A single extreme summer storm on the Isle of May resulted in wide-spread nest destruction, nesting failure and a net reduction in annual population production. While individual storms cannot be easily be attributed to climate change, it is generally believed that severe storms are becoming more common

4 - Recent declines in shag populations because of high adult mortality are most likely because of increasingly severe winter storms.

Heubeck, M., et al. "Population and breeding dynamics of European Shags Phalacrocorax aristotelis at three major colonies in Shetland, 2001-15." **Seabird 28 (2015): 55-77**. Populations in the Shetlands have markedly declined, likely due to high mortality from winter storms. While individual extreme weather events are difficult to attribute to climate change, the frequency and severity of extreme weather events is likely increasing.

5 - Shags breed later as winters have become colder

Lorentsen, S.-H., et al. "Forage fish abundance is a predictor of timing of breeding and hatching brood size in a coastal seabird." **Marine Ecology Progress Series 519 (2015): 209-220**. Median hatch date became later over the study period (1989–2009). This was partially explained by a decline in the wNAO index, indicating colder conditions. This relationship was only observed at one of two studied colonies (Røst). It is unclear if or how this change affects the population

Great Cormorant

1 - Cormorants that migrate to coastal areas during the winter are now migrating later, most likely due to less and later ice on freshwater feeding areas.

Trella, M., and Wołos, A.. "Opinions of Owners and Managers of Fishing Entities in Central and Eastern Europe on the Impact of Climate Change on Lake Fisheries Management." Fisheries & Aquatic Life 29.4 (2021): 189-201. In Poland, observational surveys show that cormorants stay later in freshwater habitats during winter than previously, due to shorter ice-cover duration. Populations are shifting to relying more heavily on freshwater, and away from coastal and marine habitats.

2 - Cormorants are expanding their range due to increased availability of prey, in large part due to declines in competing marine predators, which in turn are partially driven by climate change.

van Eerden, M. R., et al. "Expanding East: Great Cormorants Phalacrocorax carbo Thriving in the Eastern Baltic and Gulf of Finland." Ardea 109.3 (2022): 313-326. Large marine predators have declined in part due to climate change (in particular due to warmer waters and increased eutrophication), along with several other major factors, which has resulted in an increase in smaller prey. This is likely a major factor contributing to major growth in cormorant populations in southern Finland, Estonia, Lithuania and across the eastern Baltic area, which are now breeding in areas they were never historically associated with.

1.1.2 Change in European range size between present day and 2100:

Using a species distribution model (SDM) we correlated species occurrence during the breeding season with a number of terrestrial and marine environmental variables. Species range data came from the European Breeding Bird Atlas (EBBA2) database. Present-day and 2100 terrestrial data were downloaded from the WorldClim database. We used data from the MRI-ESM2 general circulation model (GCM), which is a high-performing model over Europe. Present-day and 2100 marine data were downloaded from the Bio-Oracle database which averages predictions of marine variables from several different atmospheric-oceanic general circulation models (AOGCMS; for full details see Assis et al., 2017). For the map presented in the summary we used representative concentration pathway (RCP) 4.5, which is an "intermediate" emissions scenario. All data were at 5-minute resolution.

For European shag, and northern gannet we included the following terrestrial variables: Mean temperature of the warmest month, precipitation during breeding season, isolation of landmass, area of landmass, distance to sea.

For great cormorant we included the following terrestrial variables: mean temperature of the warmest month, precipitation during breeding season, distance to sea.

For European shag, and northern gannet we included the following marine variables: sea surface temperature (during the winter), salinity, maximum chlorophyll concentration, bathymetry (depth and variance).

Several other variables may strongly influence the distribution of gannets and cormorants and it is not possible to include all possible variables in a given model. However the following variables have previously been found to be important to predicting the distribution of gannets and cormorants in Europe: average wind speed during breeding season, sea surface height, seabed substrate, average wind speed during breeding season, presence of stable ocean fronts (or bathymetric proxy)distance to fresh water, freshwater depth, freshwater ph, freshwater chlorophyll concentration, land "roughness" index. For local assessments of climate change, we recommend these variables are strongly considered. We hope to incorporate these variables into future versions of this guidance resource.

After running our model we generated a present-day map where every grid-cell is given a habitat suitability score between 0 and 1, where 1 is very suitable habitat and 0 is not at all suitable. We then compared this with a corresponding map built with 2100 data, and highlighted currently inhabited areas where 1) suitability drops sharply (i.e. by more than 0.1) and 2) suitability drops below a probability threshold set by the model. Conversely we also highlighted areas where suitability rose sharply and above a given threshold. While a drop in habitat suitability is likely to result in population declines, it is not a certainty, and it does not mean that a population will be extinct in 2100 or that a population is doomed to extinction. With conservation action and careful management, along with changes in human behaviour, such declines may be mitigated or in some cases prevented. For a full explanation of the model see the accompanying 'Methodology' folder in Appendix 2.

Underlying data were downloaded from:

Keller, V., et al. "European Breeding Bird Atlas 2: Distribution, Abundance and Change." European Bird Census Council & Lynx Edicions, Barcelona (2020). Source of range data

Fick, S. E., and Hijmans, R. J. "WorldClim 2: new 1-km spatial resolution climate surfaces for global land areas." International Journal of Climatology 37.12 (2017): 4302-4315. Source of present-day and 2100 terrestrial data.

Assis, J., et al. "Bio-ORACLE v2. 0: Extending marine data layers for bioclimatic modelling." Global Ecology and Biogeography 27.3 (2018): 277-284. Source of present-day and 2100 marine data

1.1.3 Changes in key prey species:

We first identified the key prey species for each species. This can be variable across a species' range, but if available evidence suggested at least one major population is highly dependent on a particular prey species, then typically this species would be included. Lists of prey species were compiled from published sources, then verified and expanded following consultation with conservation practitioners. Afterwards we compiled current and projected maps of prey ranges to assess where key prey species may become less common in the near future. If any of the key species are predicted to vanish or drastically reduce in abundance in the current foraging range a given species, we marked this on the summary map.

We used several sources to collate range information, but for preference we used data from COPERNICUS as they include projected abundance. For species where this was not available we used habitat suitability instead. In all cases we used RCP 4.5, which is an "intermediate" emissions scenario. For species in the COPERNICUS database we used the 0.6 maximum sustainable yield parameter, which assumes international co-operation to work towards fish-stock sustainability. Our assessment is therefore relatively conservative in terms of changes in prey species.

Northern Gannet key prey species: herring (*Clupea harengus*), saithe (*Pollachius virens*), mackerel (*Scomber scombrus*), sandeel species (*Ammodytes marinus* and *Ammodytes tobianus*), capelin (*Mallotus villosus*), sprat (*Sprattus sprattus)*, haddock (*Melanogrammus aeglefinus*) and garfish (*Belone belone*). This species will also take many other species where available, it will also forage fishery discards. Prey species list was compiled from:

Le Bot, T., et al. "Fishery discards do not compensate natural prey shortage in Northern gannets from the English Channel." Biological conservation 236 (2019): 375-384.

Mowbray, T. B. "Northern Gannet (Morus bassanus), version 1.0." In Birds of the World (S. M. Billerman, Editor). Cornell Lab of Ornithology, Ithaca, NY, USA (2020).

Pettex, E., et al. "Multi-scale foraging variability in Northern gannet (Morus bassanus) fuels potential foraging plasticity." Marine Biology 159.12 (2012): 2743-2756.

European Shag key prey species: sandeel species (*Ammodytes marinus* and *Ammodytes tobianus*), saith (*Pollachius virens*), cod (*Gadus morhua*), poor cod (*Trisopterus minutus*) and capelin (*Mallotus villosus*). Prey species list was compiled from:

Harris, M. P., and Wanless, S. "The diet of shags Phalacrocorax aristotelis during the chick-rearing period assessed by three methods." Bird study 40.2 (1993): 135-139.

Hillersøy, G., and Lorentsen, S-H. "Annual variation in the diet of breeding European shag (Phalacrocorax aristotelis) in Central Norway." Waterbirds 35.3 (2012): 420-429.

Lorentsen, S-H., Mattisson, J., and Christensen-Dalsgaard, S. "Reproductive success in the European shag is linked to annual variation in diet and foraging trip metrics." Marine Ecology Progress Series 619 (2019): 137-147.

Great Cormorant key prey species: sandeel species (*Ammodytes marinus* and Ammodytes tobianus), capelin (*Mallotus villosus*), flounder (Platichthys flesus), saithe (*Pollachius virens*), sea scorpion (*Taurulus bubalis*), sole (*Solea solea*), eelpout (*Zoarces viviparu*s) and sprat (*Sprattus sprattus*). This species also feeds on freshwater species, notably brown trout (*Salmo trutta*), salmon parr, European eel (*Anguilla anguilla*), roach (*Rutilus rutilus*), perch (*Perca fluviatilis*) and minnow (*Phoxinus phoxinus*), but these were not included in this analysis. Prey species list was compiled from:

Hatch, J. J., et al. "Great Cormorant (Phalacrocorax carbo), version 1.0." In Birds of the World (S. M. Billerman, Editor). Cornell Lab of Ornithology, Ithaca, NY, USA (2020).

van Eerden, M. R., et al. "Expanding East: Great Cormorants Phalacrocorax carbo Thriving in the Eastern Baltic and Gulf of Finland." Ardea 109.3 (2022): 313-326.

Lehikoinen, A., Heikinheimo, O., and Lappalainen, A. "Temporal changes in the diet of great cormorant (Phalacrocorax carbo sinensis) on the southern coast of Finland-comparison with available fish data." Boreal environment research 16 (suppl. B) (2011): 61–70.

Prey range information for all species were compiled from:

> Kesner-Reyes, K., et al. "AquaMaps: Predicted range maps for aquatic species." In FishBase: R. Froese & D. Pauly (Eds.) (2019). Available at: https://www.aquamaps.org

> Sailley, S., et al. "Fish abundance and catch data for the Northwest European Shelf and Mediterranean Sea from 2006 to 2098 derived from climate projections". Copernicus Climate Change Service (C3S) Climate Data Store (CDS) (2021). https://doi.org/10.24381/cds.39c97304.

1.1.4 Climate change impacts outside of Europe
Northern Gannet

Marine heatwaves in North America have resulted in wide-spread breeding failure and in some cases temporary desertion of colonies. Most likely because of prey shortages, but heat stress could play a role as well. It is difficult to attribute individual climate events to climate change, but heatwaves are becoming more common and more extreme, and will likely continue to do so. Lack of key prey species (mackerel) due to warmer average marine temperatures and over-exploitation has caused low breeding success in a southern population of gannets in Canada.

> Franci, C. D., et al. "Nutritional stress in Northern gannets during an unprecedented low reproductive success year: Can extreme sea surface temperature event and dietary change be the cause?." Comparative Biochemistry and Physiology Part A: Molecular & Integrative Physiology 181 (2015): 1-8

> d'Entremont, K. J. N., et al. "Northern Gannets (Morus bassanus) breeding at their southern limit struggle with prey shortages as a result of warming waters." ICES Journal of Marine Science 79.1 (2022): 50-60.

> Montevecchi, W. A., et al. "Ocean heat wave induces breeding failure at the southern breeding limit of the Northern Gannet Morus bassanus." Marine Ornithology 49 (2021): 71-78.

Great Cormorant

Cormorants in Greenland have spread their summer range further north, most likely due to warmer sea temperatures and changes in food availability. However, this has also likely increased the costs of migration, as cormorants have further to travel to reach ice-free areas in winter.

> White, C. R., et al. "Energetic constraints may limit the capacity of visually

guided predators to respond to Arctic warming." Journal of Zoology 289.2 (2013): 119-126.

1.2 Sensitivity (references)

We used a list of candidate traits based on that in Foden & Young (2016) that indicate high sensitivity and identified which, if any, gannets and cormorants possessed. In brief, we consulted published literature as well as expert knowledge and online databases such as Birdlife (http://datazone.birdlife.org/) and Birds of the World (https://birdsoftheworld.org), to assess whether gannets and cormorants have either 1) Specialised habitat and/or microhabitat requirement 2) Environmental tolerances or thresholds (at any life stage) that are likely to be exceeded due to climate change 3) Dependence on environmental triggers that are likely to be disrupted by climate change, 4) Dependence on interspecific interactions that are likely to be disrupted by climate change or 5) High rarity. For more detail and a full list of traits see:

> Foden, W. B. and Young, B. E. (eds.). "IUCN SSC Guidelines for Assessing Species' Vulnerability to Climate Change. Version 1.0." Occasional Paper of the IUCN Species Survival Commission No. 59 (2016). Cambridge, UK and Gland, Switzerland: IUCN Species Survival Commission. x+114pp.

1.3 Adaptive capacity (references)

We used a list of candidate traits based on that in Foden & Young (2016) that indicate adaptive capacity and identified which, if any, gannets and cormorants possessed. In brief, we consulted published literature as well as expert knowledge and online databases such as Birdlife (http://datazone.birdlife.org/) and Birds of the World (https://birdsoftheworld.org), to assess whether gannets and cormorants have either: 1) High phenotypic plasticity. 2) High dispersal ability or 3) High evolvability.

For more detail and a full list of traits see:

> Foden, W. B. and Young, B. E. (eds.). "IUCN SSC Guidelines for Assessing Species' Vulnerability to Climate Change. Version 1.0." Occasional Paper of the IUCN Species Survival Commission No. 59 (2016). Cambridge, UK and Gland, Switzerland: IUCN Species Survival Commission. x+114pp.

© Seppo Häkkinen

Appendix 1: Gulls
Sources and references for vulnerability assessment

1.1 Evidence for exposure (references)

1.1.1 Current impacts attributed to climate change:

European Herring Gull

1 - Changes in mercury cycling (due to increased sea temperatures) has led to increased exposure to mercury, with negative impacts on herring gull health

Bełdowska, M., et al. "Mercury concentration in the coastal zone of the Gulf of Gdansk as a function of changing climate—preliminary results." Baltic Sea Science Congress: new horizons for Baltic Sea science. Klaipeda University CORPI, Klaipeda 140 (2013). This paper does not specifically address bioaccumulation in gulls, but provides supporting information for the other references for this impact. Overview of how climate change can lead to changes in mercury concentration in Baltic.

Beldowska, M., et al. "Macrophyta as a vector of contemporary and historical mercury from the marine environment to the trophic web." Environmental Science and Pollution Research 22.7 (2015): 5228-5240. This paper does not specifically address bioaccumulation in gulls, but provides supporting information for the other references for this impact. Overview of how mercury concentration in the Baltic cycles into seabird tissues and how it affects ecosystem health

Saniewska, D., et al. "Climate change and its impact on the mercury cycling in the southern Baltic Sea." Ecosystem dynamics in the Baltic Sea in a climate change perspective, 03.2015. Conference Umeå, Sweden, (2015). Climate change is likely leading to increased mercury concentration cycling in the southern Baltic, this is leading to higher concentrations of mercury in herring gull tissues. While this has not been shown to be affecting the population in the Baltic, it is known that mercury poisoning can severely affect herring gull health.

Lesser Black-backed Gull

1 - Increased prey availability during the breeding season has led to population growth

Luczak, C., et al. "North Sea ecosystem change from swimming crabs to seagulls." Biology letters 8.5 (2012): 821-824. Increased numbers of swimming crabs significantly correlate with sea surface temperature increases and changes in the abundance of lesser black-backed gulls at 21 major North Sea breeding colonies (across northern France and Belgium). Though note there is some debate on whether an increase in crabs is actually the cause of the population increase.

Glaucous Gull

1 - There has been increased predation by polar bears, most likely due to reduction in sea ice and therefore a lack of alternative prey. In some years this has severely affected breeding success.

> **Prop, J., et al. "Climate change and the increasing impact of polar bears on bird populations." Frontiers in Ecology and Evolution 3 (2015): 33.** Polar bears are swapping prey due to lack of sea ice and prey more heavily on glaucous gulls (amongst other seabirds). In some years it severely affects reproductive success in Svalbard and Greenland.

2 - Climate change is likely contributing to higher concentrations of contaminents ingested by glaucous gulls. The overall effect on the population is unknown, but presumably negative.

> **Alava, J. J., et al. "Climate change–contaminant interactions in marine food webs: Toward a conceptual framework." Global Change Biology 23.10 (2017): 3984-4001.** Potential exacerbation of POPs and mercury in marine food webs due to climate change (i.e., increasing temperatures). Glaucous gulls appear to have high accumulation of compounds, but there's no established negative effect on populations. Study was conducted across their range, especially around Greenland.

3 - Climate change has contributed to a range shift in several helminth parasites, which has led to glaucous gulls being exposed to novel parasites, as well as increased parasite load. Effect on population is unknown, but presumably negative

> **Galaktionov, K. V. "Patterns and processes influencing helminth parasites of Arctic coastal communities during climate change." Journal of Helminthology 91.4 (2017): 387-408.** Review of helminth parasites across the Arctic, especially in seabirds. Notes that several parasites have been recorded in species they have never been associated with before. Most likely because boreal crustaceans are shifting north, along with associated parasites. Notable species are glaucous gulls and black-legged kittiwakes which now have significant new parasites at rapidly increasing loads.

Great Black-backed Gull

1 - Higher sea temperatures correlate with lower breeding success. Mechanism unknown, but likely mediated through prey availability

Burthe, S. J., et al. "Assessing the vulnerability of the marine bird community in the western North Sea to climate change and other anthropogenic impacts." Marine Ecology Progress Series 507 (2014): 277-295. Greater black-backed gull productivity decreases as sea surface temperature gets higher. Probably due to prey availability, study focusses on seabirds in Forth and Tay region.

Ivory Gull

1 - Ivory gulls are heavily reliant on sea ice for breeding and hunting, recent decreases in sea ice are leading to rapid changes in population size and range.

Gilg, O., et al. "Living on the edge of a shrinking habitat: the ivory gull, Pagophila eburnea, an endangered sea-ice specialist." Biology letters 12.11 (2016): 20160277. Demonstrates how closely linked Ivory gull ranges are to sea ice availability across its range, including Greenland, Svalbard, Russia and Canada. Also notes that their range is shrinking due to changes in sea ice.

2 - As a secondary impact of sea ice loss, Ivory gulls face more competition from other Ivory gulls and from other species for resources.

Hamilton, C. D., et al. "Spatial overlap among an Arctic predator, prey and scavenger in the marginal ice zone." Marine Ecology Progress Series 573 (2017): 45-59. Decreasing area of sea ice mean less area for ivory gulls, polar bears and ringed seals (both of which commonly scavenge in similar areas). This leads to increased intra- and inter- specific competition.

Black-legged Kittiwake

1 - Decreased prey availability due to warmer seas has led to lower breeding success

Sandvik, H., et al. "The decline of Norwegian kittiwake populations: modelling the role of ocean warming." Climate Research 60.2 (2014): 91-102. The study find a correlation between lower breeding success and sea surface temperature along Norwegian coast. The suggested mechanism is through prey availability as this is known to affect breeding success in various parts of their range.

2 - Kittiwake diet has changed significantly due to climate-change driven shift in prey assemblage. However, so far this has not resulted in any demonstrated change in breeding success.

Vihtakari, M., et al. "Black-legged kittiwakes as messengers of Atlantification in the Arctic." Scientific Reports 8.1 (2018): 1-11. Kittiwake diet in Svalbard changed significantly over a 10-year period (2006-2016), from predominantly Arctic species to species more associated with warmer Atlantic waters, in correlation with increase in ocean temperature and loss of sea ice. There were, however, no significant changes in clutch size or breeding success during this time.

3 - Kittiwake populations have shifted their range in response to changes in distribution of key prey species.

> **Garðarsson, A., Guðmundsson, G. A., and Lilliendahl, K. "Framvinda íslenskra ritubyggða." Bliki 32 (2013): 1-10.** Kittiwakes have redistributed across Iceland. Populations in the north have decreased, and populations in the west have increased. The population in general is relatively stable. The authors hypothesise this is likely due to redistribution of capelin, a key prey species of kittiwakes in Iceland.

4 - Climate change has contributed to a range shift in several helminth parasites, which has led to kittiwakes being exposed to novel parasites, as well as increased parasite load. Effect on population is unknown, but most likely negative

> **Galaktionov, K. V. "Patterns and processes influencing helminth parasites of Arctic coastal communities during climate change." Journal of Helminthology 91.4 (2017): 387-408.** Review of helminth parasites across the Arctic, especially in seabirds. Notes that several parasites have been recorded in species they have never been associated with before. Most likely because boreal crustaceans are shifting north, along with associated parasites. Notable species are glaucous gulls and black-legged kittiwakes which now have significant new parasites at rapidly increasing loads.

5 - Higher sea temperatures correlate with lower breeding success. Mechanism unknown, but potentially mediated through prey availability. Alternative theories suggest fishery pressure has been a large contributing factor.

> **Burthe, S. J., et al. "Assessing the vulnerability of the marine bird community in the western North Sea to climate change and other anthropogenic impacts." Marine Ecology Progress Series 507 (2014): 277-295.** Kittiwake productivity has decreased as sea surface temperature has increased. Probably due to prey availability, study focusses on seabirds in Forth and Tay region.

> **Carroll, M. J., et al. "Effects of sea temperature and stratification changes on seabird breeding success." Climate Research 66.1 (2015): 75-89.** Across 11 colonies in the UK and Ireland, kittiwake breeding success was lower in years with higher spring sea surface temperatures (and strong ocean stratification in the preceding winter). Both variables increased over the study period. However, key drivers of productivity varied between colonies. Environmental variables

were taken from the observed foraging areas of the kittiwakes. The study suggests a food-web-based mechanism for impact.

Frederiksen, M., et al. "Regional and annual variation in black-legged kittiwake breeding productivity is related to sea surface temperature." Marine Ecology Progress Series 350 (2007): 137-143. Over the study period (mid-1980s to mid-2000s, February/March sea surface temperature increased around six of six colonies in the UK and Ireland – although only significantly so in two. Breeding productivity negatively related to SST across all colonies, and for two of six individual colonies. The study suggests the mechanism is unclear, but likely operates through effects on sandeels.

6 - Kittiwake colonies have declined during periods of rapid ocean warming. Mechanism unknown, but likely due to rapid changes in marine ecosystems and prey availability

Descamps, S., et al. "Circumpolar dynamics of a marine top-predator track ocean warming rates." Global Change Biology 23.9 (2017): 3770-3780. Rapid increases in SST around circumpolar kittiwake colonies in the 1990s conicided with steep declines in colony size. The suggested mechanism is abrupt "regime shifts" that changed prey availability.

7 - Extreme storms during the non-breeding season have led to mass mortality of kittiwakes ('wrecks')

Clairbaux, M., et al. "North Atlantic winter cyclones starve seabirds." Current Biology 31.17 (2021): 3964-3971. Following heavy storm action, seabird mortality increases due to increased difficulty foraging (rather than increased energetic costs). The authors use a multi-species dataset (puffins, little auks, common murres, and thick-billed murres) over a wide area of the Atlantic basin. They conclude that seabirds around Iceland and the Barents Sea (along with several N. American sites) are particularly vulnerable. Climate change is likely to be a contributing factor to present and future storm mortality.

8 - Extreme storms during the kittiwake breeding season have led to wide-spread nest destruction, nesting failure and a net reduction in annual population production

Newell, M., et al. "Effects of an extreme weather event on seabird breeding success at a North Sea colony." Marine Ecology Progress Series 532 (2015): 257-268. A single extreme summer storm on the Isle of May resulted in wide-spread nest destruction, nesting failure and a net reduction in annual population production. While individual storms cannot be easily be attributed to climate change, it is generally believed that severe storms are becoming more common.

Yellow-legged Gull

1 - Changes in prey availability have caused species to swap prey

Calado, J. G., et al. "Anthropogenic food resources, sardine decline and environmental conditions have triggered a dietary shift of an opportunistic seabird over the last 30 years on the northwest coast of Spain." Regional Environmental Change 20.1 (2020): 1-13. Yellow-legged gulls have changed their diet following declines in sardine populations, driven by climate change and fishery activity. Study based on several populations on the north-west coast of Spain.

1.1.2 Change in European range size between present day and 2100:

Using a species distribution model (SDM) we correlated species occurrence during the breeding season with a number of terrestrial and marine environmental variables. Species range data came from the European Breeding Bird Atlas (EBBA2) database. Present-day and 2100 terrestrial data were downloaded from the WorldClim database. We used data from the MRI-ESM2 general circulation model (GCM), which is a high-performing model over Europe. Present-day and 2100 marine data were downloaded from the Bio-Oracle database which averages predictions of marine variables from several different atmospheric-oceanic general circulation models (AOGCMS; for full details see Assis et al., 2017). For the map presented in the summary we used representative concentration pathway (RCP) 4.5, which is an "intermediate" emissions scenario. All data were at 5-minute resolution.

For European herring gull, Audouin's gull, lesser black-backed gull, glaucous gull, great black-backed gull, ivory gull, black-legged kittiwake, and Sabine's gull we included the following terrestrial variables: Mean temperature of the warmest month, precipitation during breeding season, isolation of landmass, area of landmass, distance to sea

For Caspian gull, and yellow-legged gull we included the following terrestrial variables: mean temperature of the warmest month, precipitation during breeding season, distance to sea.

For Audouin's gull, lesser black-backed gull, glaucous gull, great black-backed gull, ivory gull, black-legged kittiwake, and Sabine's gull we included the following marine variables: sea surface temperature (during the winter), salinity, maximum chlorophyll concentration, bathymetry (depth and variance).

Several other variables may strongly influence the distribution of gulls and it is not possible to include all possible variables in a given model. However the following variables have previously been found to be important to predicting the distribution of gulls in Europe: sea level height. For local assessments of climate change, we recommend these variables are strongly considered. We hope to incorporate these variables into future versions of this guidance resource.

After running our model we generated a present-day map where every grid-cell is given a habitat suitability score between 0 and 1, where 1 is very suitable habitat and 0 is not at all suitable. We then compared this with a corresponding map built with 2100 data, and highlighted currently inhabited areas where 1) suitability drops sharply (i.e. by more than 0.1) and 2) suitability drops below a probability threshold set by the model. Conversely we also highlighted areas where suitability rose sharply and above a given threshold. While a drop in habitat suitability is likely to result in population declines, it is not a certainty, and it does not mean that a population will be extinct in 2100 or that a population is doomed to extinction. With conservation action and careful management, along with changes in human behaviour, such declines may be mitigated or in some cases prevented. For a full explanation of the model see the accompanying 'Methodology' folder in Appendix 2.

Underlying data were downloaded from:

Keller, V., et al. "European Breeding Bird Atlas 2: Distribution, Abundance and Change." European Bird Census Council & Lynx Edicions, Barcelona (2020). Source of range data

Fick, S. E., and Hijmans, R. J. "WorldClim 2: new 1-km spatial resolution climate surfaces for global land areas." International Journal of Climatology 37.12 (2017): 4302-4315. Source of present-day and 2100 terrestrial data.

Assis, J., et al. "Bio-ORACLE v2. 0: Extending marine data layers for bioclimatic modelling." Global Ecology and Biogeography 27.3 (2018): 277-284. Source of present-day and 2100 marine data

1.1.3 Changes in key prey species:
We first identified the key prey species for each species. This can be variable across a species' range, but if available evidence suggested at least one major population is highly dependent on a particular prey species, then typically this species would be included. Lists of prey species were compiled from published sources, then verified and expanded following consultation with conservation practitioners. Afterwards we compiled current and projected maps of prey ranges to assess where key prey species may become less common in the near future. If any of the key species are predicted to vanish or drastically reduce in abundance in the

current foraging range a given species, we marked this on the summary map. We used several sources to collate range information, but for preference we used data from COPERNICUS as they include projected abundance. For species where this was not available we used habitat suitability instead. In all cases we used RCP 4.5, which is an "intermediate" emissions scenario. For species in the COPERNICUS database we used the 0.6 maximum sustainable yield parameter, which assumes international co-operation to work towards fish-stock sustainability. Our assessment is therefore relatively conservative in terms of changes in prey species.

European Herring Gull key prey species: This species has a very varied diet, including fish, discards, bivalves, gastropods, crustaceans, squid, insects, refuse and other seabirds, among many other food sources. No key species were identified, so the key prey assessment is not complete

Audouin's Gull key prey species: sardines (*Sardina pilchardus*), Atlantic saury (*Scomberesox saurus*) and blue whiting (*Micromesistius poutassou*). This species is also known to feed on various other demersal and pelagic fish, as well as various terrestrial food sources, but no other species could be firmly identified as key prey species. Prey species list was compiled from:

Burger, J., et al. "Audouin's Gull (Ichthyaetus audouinii), version 1.0." In Birds of the World (J. del Hoyo, A. Elliott, J. Sargatal, D. A. Christie, and E. de Juana, Editors). Cornell Lab of Ornithology, Ithaca, NY, USA (2020)

Calado, J. G., et al. "Seasonal and annual differences in the foraging ecology of two gull species breeding in sympatry and their use of fishery discards." Journal of Avian Biology 49.1 (2018).

Caspian Gull key prey species: This species has a varied diet of invertebrates. No key species were identified and so the key prey assessment is not complete

Lesser Black-backed Gull key prey species: herring (*Clupea harengus*), saithe (*Pollachius virens*) and sandeel species (*Ammodytes marinus* and *Ammodytes tobianus*). Many populations also rely heavily on discards, refuse and egg/chick predation. These food sources were not included in this assessment. Prey species list was compiled from:

Burger, J., et al. "Lesser Black-backed Gull (Larus fuscus), version 1.0." In Birds of the World (J. del Hoyo, A. Elliott, J. Sargatal, D. A. Christie, and E. de Juana, Editors). Cornell Lab of Ornithology, Ithaca, NY, USA (2020).

Bustnes, J. O., Barrett, R. T., and Helberg, M. "Northern Lesser Black-Backed Gulls: What do They Eat?." Waterbirds 33.4 (2010): 534-540.

Glaucous Gull key prey species: Arctic cod (*Boreogadus saida*), cod (*Gadus morhua*), sandeel species (*Ammodytes marinus* and *Ammodytes tobianus*), capelin (*Mallotus villosus*) and herring (*Clupea harengus*). Prey species list was compiled from:

Weiser, E. and Gilchrist, H. G. "Glaucous Gull (Larus hyperboreus), version 1.0." In Birds of the World (S. M. Billerman, Editor). Cornell Lab of Ornithology, Ithaca, NY, USA (2020).

Great Black-backed Gull key prey species: capelin (*Mallotus villosus*), cod (*Gadus morhua*), mackerel (*Scomber scombrus*) and herring (*Clupea harengus*). Some populations also rely on other seabirds (including eggs, chicks and adults) as an important prey source. These were not included in the key prey assessment. Prey species list was compiled from:

Good, T. P. "Great Black-backed Gull (Larus marinus), version 1.0." In Birds of the World (S. M. Billerman, Editor). Cornell Lab of Ornithology, Ithaca, NY, USA (2020).

Ivory Gull key prey species: Arctic cod (*Boreogadus saida*). This species also preys frequently on invertebrates and other fish, however these were not included in the assessment. Prey species list was compiled from:

Mallory, M. L., et al. "Ivory Gull (Pagophila eburnea), version 1.0." In Birds of the World (S. M. Billerman, Editor). Cornell Lab of Ornithology, Ithaca, NY, USA (2020).

Black-legged Kittiwake key prey species: sandeel species (*Ammodytes marinus* and *Ammodytes tobianus*), sprat (*Sprattus sprattus*), Arctic cod (*Boreogadus saida*), capelin (*Mallotus villosus*) and herring (*Clupea harengus*). Prey species list was compiled from:

Hatch, S. A., Robertson, G. J., and Baird, P. H. "Black-legged Kittiwake (Rissa tridactyla), version 1.0." In Birds of the World (S. M. Billerman, Editor). Cornell Lab of Ornithology, Ithaca, NY, USA (2020).

Fauchald, Per, et al. "The status and trends of seabirds breeding in Norway and Svalbard." NINA rapport 1151. Norsk institutt for naturforskning (2015): 1-84.

Johansen M., et al. "International Black-legged Kittiwake Conservation Strategy and Action Plan" Circumpolar Seabird Expert Group. Conservation of Arctic Flora and Fauna, Akureyri, Iceland (2020).

Sabine's Gull key prey species: This species has a varied diet of invertebrates and fish. No key species were identified and so the key prey assessment is not complete

Yellow-legged Gull key prey species: Henslow's swimming crab (*Polybius henslowii*), sardines (*Sardina pilchardus*), chub mackeral (*Scomber colias*), blue whiting (*Micromesistius poutassou*) and hake (*Merluccius merluccius*). This species has a varied diet of pelagic and demersal fish, as well as several terrestrial food sources. Note that several of the fish species here may be from fishery discards. Prey species list was compiled from:

> del Hoyo, J., et al. "Yellow-legged Gull (Larus michahellis), version 1.0." In Birds of the World (J. del Hoyo, A. Elliott, J. Sargatal, D. A. Christie, and E. de Juana, Editors). Cornell Lab of Ornithology, Ithaca, NY, USA (2020).

Prey range information for all species were compiled from:

> Kesner-Reyes, K., et al. "AquaMaps: Predicted range maps for aquatic species." In FishBase: R. Froese & D. Pauly (Eds.) (2019). Available at: https://www.aquamaps.org

> Sailley, S., et al. "Fish abundance and catch data for the Northwest European Shelf and Mediterranean Sea from 2006 to 2098 derived from climate projections". Copernicus Climate Change Service (C3S) Climate Data Store (CDS) (2021). https://doi.org/10.24381/cds.39c97304.

1.1.4 Climate change impacts outside of Europe
European Herring Gull

Increased flooding due to sea level rise has led to the reduction or destruction of several populations in the US.

> Burger, J., and Gochfeld, M. "Habitat, population dynamics, and metal levels in colonial waterbirds: a food chain approach." CRC Pres, New York (2016).

Glaucous Gull

Glaucous gull colonies display higher rates of cannabalism and lower breeding success in response to higher sea temperatures. This is presumably due to lack of marine prey, and is likely to be excerabated with further climate change.

> Hayward, J. L., et al. "Egg cannibalism in a gull colony increases with sea surface temperature." The Condor 116.1 (2014): 62-73.

Ivory Gull

Climate change is known to have several other impacts in other parts of the species range, in particular through changing winter food supplies, increasing competition with other marine birds, and increased predation due to increased access to previously isolated colonies.

> Gilchrist, H. G., and Mallory, M. L. "Declines in abundance and distribution of the ivory gull (Pagophila eburnea) in Arctic Canada." Biological Conservation 121.2 (2005): 303-309.

> Hamilton, C. D., et al. "Spatial overlap among an Arctic predator, prey and scavenger in the marginal ice zone." Marine Ecology Progress Series 573 (2017): 45-59.

> Yannic, G., et al. "Complete breeding failures in ivory gull following unusual rainy storms in North Greenland." Polar Research 33.1 (2014): 22749.

Black-legged Kittiwake

Recent heatwaves in the North Pacific have resulted in mass mortality and widespread breeding failure at kittiwake colonies

> Johansen M., et al. "International Black-legged Kittiwake Conservation Strategy and Action Plan" Circumpolar Seabird Expert Group. Conservation of Arctic Flora and Fauna, Akureyri, Iceland (2020). Note: data not provided in report.

1.2 Sensitivity (references)

We used a list of candidate traits based on that in Foden & Young (2016) that indicate high sensitivity and identified which, if any, gulls possessed. In brief, we consulted published literature as well as expert knowledge and online databases such as Birdlife (http://datazone.birdlife.org/) and Birds of the World (https://birdsoftheworld.org), to assess whether gulls have either 1) Specialised habitat and/or microhabitat requirement 2) Environmental tolerances or thresholds (at any life stage) that are likely to be exceeded due to climate change 3) Dependence on environmental triggers that are likely to be disrupted by climate change, 4) Dependence on interspecific interactions that are likely to be disrupted by climate change or 5) High rarity.

For more detail and a full list of traits see:

> Foden, W. B. and Young, B. E. (eds.). "IUCN SSC Guidelines for Assessing Species' Vulnerability to Climate Change. Version 1.0." Occasional Paper of the IUCN Species Survival Commission No. 59 (2016). Cambridge, UK and Gland, Switzerland: IUCN Species Survival Commission. x+114pp.

1.3 Adaptive capacity (references)

We used a list of candidate traits based on that in Foden & Young (2016) that indicate adaptive capacity and identified which, if any, gulls possessed. In brief, we consulted published literature as well as expert knowledge and online databases such as Birdlife (http://datazone.birdlife.org/) and Birds of the World (https://birdsoftheworld.org), to assess whether gulls have either: 1) High phenotypic plasticity. 2) High dispersal ability or 3) High evolvability.

For more detail and a full list of traits see:

Foden, W. B. and Young, B. E. (eds.). "IUCN SSC Guidelines for Assessing Species' Vulnerability to Climate Change. Version 1.0." Occasional Paper of the IUCN Species Survival Commission No. 59 (2016). Cambridge, UK and Gland, Switzerland: IUCN Species Survival Commission. x+114pp.

© Seppo Häkkinen

Appendix 1: Loons/Divers and Grebes
Sources and references for vulnerability assessment

1.1 Evidence for exposure (references)

1.1.1 Current impacts attributed to climate change:

No impacts were recorded for Loons/Divers and Grebes in Europe

1.1.2 Change in European range size between present day and 2100:

Using a species distribution model (SDM) we correlated species occurrence during the breeding season with a number of terrestrial and marine environmental variables. Species range data came from the European Breeding Bird Atlas (EBBA2) database. Present-day and 2100 terrestrial data were downloaded from the WorldClim database. We used data from the MRI-ESM2 general circulation model (GCM), which is a high-performing model over Europe. Present-day and 2100 marine data were downloaded from the Bio-Oracle database which averages predictions of marine variables from several different atmospheric-oceanic general circulation models (AOGCMS; for full details see Assis et al., 2017). For the map presented in the summary we used representative concentration pathway (RCP) 4.5, which is an "intermediate" emissions scenario. All data were at 5-minute resolution.

For Arctic loon, common loon, horned grebe, and red-necked grebe we included the following terrestrial variables: Mean temperature of the warmest month, precipitation during breeding season, distance to sea

For red-throated loon we included the following terrestrial variables: Mean temperature of the warmest month, precipitation during breeding season, isolation of landmass, area of landmass, distance to sea

No marine variables were included for this species group, as they are predominantly terrestrial during the breeding season.

Several other variables may strongly influence the distribution of loons, divers and grebes and it is not possible to include all possible variables in a given model. However the following variables have previously been found to be important to predicting the distribution of loons, divers and grebes in Europe: distance to fresh water, freshwater depth, freshwater ph, freshwater chlorophyll concentration, land "roughness" index. For local assessments of climate change, we recommend these variables are strongly considered. We hope to incorporate these variables into future versions of this guidance resource.

After running our model we generated a present-day map where every grid-cell is given a habitat suitability score between 0 and 1, where 1 is very suitable habitat and 0 is not at all suitable. We then compared this with a corresponding map built with 2100 data, and highlighted currently inhabited areas where 1) suitability drops sharply (i.e. by more than 0.1) and 2) suitability drops below a probability threshold set by the model. Conversely we also highlighted areas where suitability rose sharply and above a given threshold. While a drop in habitat suitability is likely to result in population declines, it is not a certainty, and it does not mean that a population will be extinct in 2100 or that a population is doomed to extinction. With conservation action and careful management, along with changes in human behaviour, such declines may be mitigated or in some cases prevented. For a full explanation of the model see the accompanying 'Methodology' folder in Appendix 2.

Underlying data were downloaded from:

Keller, V., et al. "European Breeding Bird Atlas 2: Distribution, Abundance and Change." European Bird Census Council & Lynx Edicions, Barcelona (2020). Source of range data

Fick, S. E., and Hijmans, R. J. "WorldClim 2: new 1-km spatial resolution climate surfaces for global land areas." International Journal of Climatology 37.12 (2017): 4302-4315. Source of present-day and 2100 terrestrial data.

Assis, J., et al. "Bio-ORACLE v2. 0: Extending marine data layers for bioclimatic modelling." Global Ecology and Biogeography 27.3 (2018): 277-284. Source of present-day and 2100 marine data

1.1.3 Changes in key prey species:

We first identified the key prey species for each species. This can be variable across a species' range, but if available evidence suggested at least one major population is highly dependent on a particular prey species, then typically this species would be included. Lists of prey species were compiled from published sources, then verified and expanded following consultation with conservation practitioners. Afterwards we compiled current and projected maps of prey ranges to assess where key prey species may become less common in the near future. If any of the key species are predicted to vanish or drastically reduce in abundance in the current foraging range a given species, we marked this on the summary map.

We used several sources to collate range information, but for preference we used data from COPERNICUS as they include projected abundance. For species where this was not available we used habitat suitability instead. In all cases we used RCP 4.5, which is an "intermediate" emissions scenario. For species in the COPERNICUS database we used the 0.6 maximum sustainable yield parameter, which assumes international co-operation to work towards fish-stock sustainability. Our assessment is therefore relatively conservative in terms of changes in prey species.

Arctic Loon key prey species: herring (*Clupea harengus*), sprat (*Sprattus sprattus*) and cod (*Gadus morhua*). This species will also prey on freshwater fish, especially during the breeding season, notably salmonids, perch, roach, minnows and sticklebacks. However, freshwater species were not included in the key prey assessment. Prey species list was compiled from:

> Jackson, D. B. "Environmental correlates of lake occupancy and chick survival of black-throated divers Gavia arctica in Scotland." Bird Study 52.3 (2005): 225-236.

> Russell, R. W. "Arctic Loon (Gavia arctica), version 1.0." In Birds of the World (S. M. Billerman, Editor). Cornell Lab of Ornithology, Ithaca, NY, USA (2020).

> Söderlund, E. Effects of whitefish speciation on piscivorous birds. A dietary study of piscivorous birds in central and northern Sweden. Umeå University (2021). MSc Thesis.

Common Loon key prey species: During the breeding season this species preys primarily on salmonids. However, freshwater species are not assessed as part of the key prey assessment. Also feeds on a wide variety of marine species, however no key prey species were identified. Currently there is no key prey assessment for this species

Red-throated Loon key prey species: herring (*Clupea harengus*), sprat (*Sprattus sprattus*), sandeels (*Ammodytes marinus*), cod (*Gadus morhua*) and smelt (*Osmerus eperlanus*). This species also commonly preys on freshwater fish, especially during the breeding season, however these were not included in the key prey assessment. Prey species list was compiled from:

> Eriksson, M. O. G., and Paltto, H. "Vattenkemi och fiskbeståndens sammansättning i storlommens Gavia arctica häckningssjöar, samt en jämförelse med smålommens Gavia stellata fiskesjöar." Ornis Svecica 20.1 (2010): 3-30.

> Rizzolo, D. J., et al. "Red-throated Loon (Gavia stellata), version 2.0." In

Birds of the World (P. G. Rodewald and B. K. Keeney, Editors). Cornell Lab of Ornithology, Ithaca, NY, USA (2020).

Horned Grebe key prey species: three-spined sticklebacks (*Gasterosteus aculeatus*) and smelt (*Osmerus eperlanus*). This species also preys on small aquatic and airborne arthropods, particularly during the breeding season (beetles, dragonflies and damselflies, mayflies etc.). Invertebrates are not currently included in key prey assessments. While this species does prey on other fish species, none were identified as key species so were not included. Prey species list was compiled from:

Dillon, I. A., Hancock, M. H. , and Summers, R. W. "Provisioning of Slavonian Grebe Podiceps auritus chicks at nests in Scotland." Bird Study 57.4 (2010): 563-567.

Piersma, T. "Body size, nutrient reserves and diet of Red-necked and Slavonian Grebes Podiceps grisegena and P. auritus on Lake IJsselmeer, The Netherlands." Bird Study 35.1 (1988): 13-24.

Sonntag, N., Garthe, S., and Adler, S. "A freshwater species wintering in a brackish environment: Habitat selection and diet of Slavonian grebes in the southern Baltic Sea." Estuarine, Coastal and Shelf Science 84.2 (2009): 186-194.

Stedman, S. J. (2020). "Horned Grebe (Podiceps auritus), version 1.0." In Birds of the World (S. M. Billerman, Editor). Cornell Lab of Ornithology, Ithaca, NY, USA (2020).

Red-necked Grebe key prey species: Smelt (*Osmerus eperlanus*), pilchard (*Gasterosteus aculeatus*), three-spined stickleback (*Crangon crangon*) and prawns. This species feeds on various fish and invertebrates, including many freshwater species (dragonfly and caddis fly larvae among many others). Freshwater species were not included in the key prey assessment. Prey species list was compiled from:

Stout, B. E. and Nuechterlein, G. L. "Red-necked Grebe (Podiceps grisegena), version 1.0." In Birds of the World (S. M. Billerman, Editor). Cornell Lab of Ornithology, Ithaca, NY, USA (2020).

Piersma, T. "Body size, nutrient reserves and diet of Red-necked and Slavonian Grebes Podiceps grisegena and P. auritus on Lake IJsselmeer, The Netherlands." Bird Study 35.1 (1988): 13-24.

Prey range information for all species were compiled from:

Kesner-Reyes, K., et al. "AquaMaps: Predicted range maps for aquatic species." In FishBase: R. Froese & D. Pauly (Eds.) (2019). Available at: https://www.aquamaps.org

Sailley, S., et al. "Fish abundance and catch data for the Northwest European Shelf and Mediterranean Sea from 2006 to 2098 derived from climate projections". Copernicus Climate Change Service (C3S) Climate Data Store (CDS) (2021). https://doi.org/10.24381/cds.39c97304.

1.1.4 Climate change impacts outside of Europe
Common Loon

Several impacts of climate change have been noted in North American populations, including decreased brood size, changes in migration patterns, increased energetic stress due to higher temperatures, and an increase in exposure to mercury.

Paruk, J. D., et al. "Common Loon (Gavia immer), version 2.0." In Birds of the World (P. G. Rodewald and B. K. Keeney, Editors). Cornell Lab of Ornithology, Ithaca, NY, USA (2021).

Bianchini, K., et al. "Drivers of declines in Common Loon (Gavia immer) productivity in Ontario, Canada." Science of the Total Environment 738 (2020): 139724.

Red-throated Loon

Red-throated Loon been caused by "red tides" (mass growth of red algae). It is difficult to attribute one event to climate change, but red tides have become more common and more widespread in California and globally, which has been linked to the effects of climate change.

Jessup, D. A., et al. "Mass stranding of marine birds caused by a surfactant-producing red tide." PLoS One 4.2 (2009): e4550.

1.2 Sensitivity (references)

We used a list of candidate traits based on that in Foden & Young (2016) that indicate high sensitivity and identified which, if any, loons, divers and grebes possessed. In brief, we consulted published literature as well as expert knowledge and online databases such as Birdlife (http://datazone.birdlife.org/) and Birds of the World (https://birdsoftheworld.org), to assess whether loons, divers and grebes have either 1) Specialised habitat and/or microhabitat requirement 2) Environmental tolerances or thresholds (at any life stage) that are likely to be exceeded due to climate change 3) Dependence on environmental triggers that are likely to be disrupted by climate change, 4) Dependence on interspecific interactions that are likely to be disrupted by climate change or 5) High rarity.

For more detail and a full list of traits see:

Foden, W. B. and Young, B. E. (eds.). "IUCN SSC Guidelines for Assessing Species' Vulnerability to Climate Change. Version 1.0." Occasional Paper of the IUCN Species Survival Commission No. 59 (2016). Cambridge, UK and Gland, Switzerland: IUCN Species Survival Commission. x+114pp.

1.3 Adaptive capacity (references)

We used a list of candidate traits based on that in Foden & Young (2016) that indicate adaptive capacity and identified which, if any, loons, divers and grebes possessed. In brief, we consulted published literature as well as expert knowledge and online databases such as Birdlife (http://datazone.birdlife.org/) and Birds of the World (https://birdsoftheworld.org), to assess whether loons, divers and grebes have either: 1) High phenotypic plasticity. 2) High dispersal ability or 3) High evolvability.

For more detail and a full list of traits see:

Foden, W. B. and Young, B. E. (eds.). "IUCN SSC Guidelines for Assessing Species' Vulnerability to Climate Change. Version 1.0." Occasional Paper of the IUCN Species Survival Commission No. 59 (2016). Cambridge, UK and Gland, Switzerland: IUCN Species Survival Commission. x+114pp.

© Seppo Häkkinen

Appendix 1: Petrels and Shearwaters
Sources and references for vulnerability assessment

1.1 Evidence for exposure (references)

1.1.1 Current impacts attributed to climate change:

Cory's Shearwater

1 - New colonies have been established outside of the species' normal range. The cause is uncertain, but likely related to prey range shifts and warming conditions.

> **Munilla, Ignacio, et al. "Colony foundation in an oceanic seabird." PloS one 11.2 (2016): e0147222.** Cory's shearwaters have established several breeding colonies in Galicia, an area they have never historically been associated with. The authors note this is a rare event, as, like most shearwaters, Cory's shearwaters have high site fidelity. The cause is unknown, but the authors hypothesis the underlying cause is generally warming condition and shift of warm-water prey species north.

Northern Fulmar

1 - Warmer winters have resulted in lower adult survival and lower reproductive success in the following year. Mechanism unknown, potentially could be related to marine productivity

> **Grosbois, V., and Thompson, P. M.. "North Atlantic climate variation influences survival in adult fulmars." Oikos 109.2 (2005): 273-290.** Adult survival decreased over the study period, in correlation with winter climate conditions (WNAO). Mechanism unknown. Study in Eynhallow, Orkneys.
> **Lewis, S., et al. "Effects of extrinsic and intrinsic factors on breeding success in a long lived seabird." Oikos 118.4 (2009): 521-528.** Confirms the former using data from the same population (Eynhallow, Orkneys). Effects of NAO (and increased winter SST) have a negative, lagged impact on fulmar breeding success.

2 - Higher sea temperatures typically correlate with lower breeding success. Mechanism unknown, but likely mediated through prey availability. Continued warming may cause long-term declines in populations.

> **Burthe, S. J., et al. "Assessing the vulnerability of the marine bird community in the western North Sea to climate change and other anthropogenic impacts." Marine Ecology Progress Series 507 (2014): 277-**

295. Fulmar productivity decreases as sea surface temperature gets higher. Probably due to prey availability, study focusses on seabirds in Forth and Tay region.

European Storm-petrel

1 - High winds and storms in the non-breeding season causes increased mortality, lower body condition, and reduced breeding success. While individual extreme climate events are difficult to attribute to climate change, it is likely that climate change is driving an increase in their frequency and/or intensity.

> **Zuberogoitia, I., et al. "Assessing the impact of extreme adverse weather on the biological traits of a European storm petrel colony." Population ecology 58.2 (2016): 303-313.** Reproductive breeding success was lower, moulting occurred later and more skipped breeding occurred in years following a winter with adverse weather. Protracted periods of continuous gale-force winds prevents petrels from feeding, and they become exhausted and severely weakened. Study was in Aketx colony (Biscay, north of Spain).

2 - A shift towards warmer, drier and calmer conditions has resulted in lower storm petrel abundance. The mechanism is unknown, but likely related to changes in marine ecosystem and key prey availability.

> **Hemery, G., et al. "Detecting the impact of oceano-climatic changes on marine ecosystems using a multivariate index: the case of the Bay of Biscay (North Atlantic-European Ocean)." Global Change Biology 14.1 (2008): 27-38.** Abundance of European storm-petrels in the Bay of Biscay declined from 1974 to 2000. The number of breeding pairs per year was negatively correlated with a local multivariate climate index (combining 11 oceanic and atmospheric variables). Storm petrels seem to have suffered locally from a trend towards warmer, drier years with calmer sea surface conditions.

Manx Shearwater

1 - Reduced prey availability during the breeding season has led to longer foraging trips and lower condition in adults and chicks

> **Riou, S., et al. "Recent impacts of anthropogenic climate change on a higher marine predator in western Britain." Marine Ecology Progress Series 422 (2011): 105-112.** In warmer years, prey is less available and adults must forage further to find prey. As a result, adults breed later and chicks reach lower peak and fledgling status. Study conducted on Skomer Island over several breeding seasons.

1.1.2 Change in European range size between present day and 2100:

Using a species distribution model (SDM) we correlated species occurrence during the breeding season with a number of terrestrial and marine environmental variables. Species range data came from the European Breeding Bird Atlas (EBBA2) database. Present-day and 2100 terrestrial data were downloaded from the WorldClim database. We used data from the MRI-ESM2 general circulation model (GCM), which is a high-performing model over Europe. Present-day and 2100 marine data were downloaded from the Bio-Oracle database which averages predictions of marine variables from several different atmospheric-oceanic general circulation models (AOGCMS; for full details see Assis et al., 2017). For the map presented in the summary we used representative concentration pathway (RCP) 4.5, which is an "intermediate" emissions scenario. All data were at 5-minute resolution.

For Cory's shearwater, northern fulmar, band-rumped storm-petrel, Leach's storm-petrel, European storm-petrel, and Manx shearwater we included the following terrestrial variables: mean temperature of the warmest month, precipitation during breeding season, isolation of landmass, area of landmass, distance to sea.

For Cory's shearwater, northern fulmar, band-rumped storm-petrel, Leach's storm-petrel, European storm-petrel, and Manx shearwater we included the following marine variables: sea surface temperature (during the winter), salinity, maximum chlorophyll concentration, bathymetry (depth and variance).

Several other variables may strongly influence the distribution of petrels and shearwaters and it is not possible to include all possible variables in a given model. However the following variables have previously been found to be important to predicting the distribution of petrels and shearwaters in Europe: average wind speed during breeding season, presence of stable ocean fronts (or bathymetric proxy). For local assessments of climate change, we recommend these variables are strongly considered. We hope to incorporate these variables into future versions of this guidance resource.

After running our model we generated a present-day map where every grid-cell is given a habitat suitability score between 0 and 1, where 1 is very suitable habitat and 0 is not at all suitable. We then compared this with a corresponding map built with 2100 data, and highlighted currently inhabited areas where 1) suitability drops sharply (i.e. by more than 0.1) and 2) suitability drops below a probability threshold set by the model. Conversely we also highlighted areas where suitability rose sharply and above a given threshold. While a drop in habitat suitability is likely to result in population declines, it is not a certainty, and it does not mean that a population will be extinct in 2100 or that a population is doomed to extinction. With conservation action and careful management, along with changes in human behaviour, such declines may be mitigated or in some cases prevented. For a full explanation of the model see the accompanying 'Methodology' folder in Appendix 2.

Underlying data were downloaded from:

Keller, V., et al. "European Breeding Bird Atlas 2: Distribution, Abundance and Change." European Bird Census Council & Lynx Edicions, Barcelona (2020). Source of range data

Fick, S. E., and Hijmans, R. J. "WorldClim 2: new 1-km spatial resolution climate surfaces for global land areas." International Journal of Climatology 37.12 (2017): 4302-4315. Source of present-day and 2100 terrestrial data.

Assis, J., et al. "Bio-ORACLE v2. 0: Extending marine data layers for bioclimatic modelling." Global Ecology and Biogeography 27.3 (2018): 277-284. Source of present-day and 2100 marine data

1.1.3 Changes in key prey species:

We first identified the key prey species for each species. This can be variable across a species' range, but if available evidence suggested at least one major population is highly dependent on a particular prey species, then typically this species would be included. Lists of prey species were compiled from published sources, then verified and expanded following consultation with conservation practitioners. Afterwards we compiled current and projected maps of prey ranges to assess where key prey species may become less common in the near future. If any of the key species are predicted to vanish or drastically reduce in abundance in the current foraging range a given species, we marked this on the summary map.

We used several sources to collate range information, but for preference we used data from COPERNICUS as they include projected abundance. For species where this was not available we used habitat suitability instead. In all cases we used RCP 4.5, which is an "intermediate" emissions scenario. For species in the COPERNICUS database we used the 0.6 maximum sustainable yield parameter, which assumes international co-operation to work towards fish-stock sustainability. Our assessment is therefore relatively conservative in terms of changes in prey species.

Cory's Shearwater key prey species: saury (*Scombresox saurus*), chub mackerel (*Scomber colias*), sardines (*Sardina pilchardus*) and garfish (*Belone belone*). This species also preys frequently on cephalopods and other fish, however these were not included in the assessment. Prey species list was compiled from:

Alonso, H., et al. "Parent–offspring dietary segregation of Cory's shearwaters breeding in contrasting environments." Marine Biology 159.6

(2012): 1197-1207.

Northern Fulmar key prey species: Atlantic cod (*Mallotus villosus*), sandeel species (*Ammodytes marinus* and *Ammodytes tobianus*), herring (*Clupea harengus*), Norway pout (*Trisopterus esmarkii*), whiting (*Merlangius merlangus*), squid (*Gonatus fabricii*) and crustaceans. This species also heavily preys on various crustacean and cephalopod species, as well as fishery discards. However currently these are not included in the key prey assessment. Prey species list was compiled from:

> Phillips, R. A., et al. "Diet of the northern fulmar Fulmarus glacialis: reliance on commercial fisheries?." Marine Biology 135.1 (1999): 159-170.

> Mallory, M. L., Hatch, S. A., And Nettleship, D. N. "Northern Fulmar (Fulmarus glacialis), version 1.0." In Birds of the World (S. M. Billerman, Editor). Cornell Lab of Ornithology, Ithaca, NY, USA (2020).

Band-rumped Storm-petrel key prey species: blue whiting (*Micromesistius poutassou*) and poor cod (*Trisopterus minutus* along with other *Trisopterus* sp.). The diet of this species is poorly characterised in Europe, it may have other key prey species that have yet to be identified, and may in addition rely on discards in some populations. Prey species list was compiled from:

> Carreiro, A. R., et al. "Metabarcoding, stables isotopes, and tracking: unraveling the trophic ecology of a winter-breeding storm petrel (Hydrobates castro) with a multimethod approach." Marine Biology 167.2 (2020): 1-13.

Leach's Storm-petrel key prey species: glacier lantern fish (*Benthosema glaciale*) and Arctic telescope (*Protomyctophum arcticum*). Prey species list was compiled from:

> Hedd, A., and Montevecchi, W. A. "Diet and trophic position of Leach's storm-petrel Oceanodroma leucorhoa during breeding and moult, inferred from stable isotope analysis of feathers." Marine Ecology Progress Series 322 (2006): 291-301.

European Storm-petrel key prey species: sprat (*Sprattus sprattus*) and sandeels (*Ammodytes marinus*). This species commonly preys on crustaceans, zooplankton and other marine invertebrates, these are a major component of diet in many populations. However, presently these species are not included in the key prey assessment. Prey species list was compiled from:

Carboneras, C., Jutglar, F., and Kirwan (2021), G. M. "European Storm-Petrel (Hydrobates pelagicus), version 1.1." In Birds of the World (Editor not available). Cornell Lab of Ornithology, Ithaca, NY, USA (2020)

D'Elbee, J., and Hemery, G. "Diet and foraging behaviour of the British Storm Petrel Hydrobates pelagicus in the Bay of Biscay during summer." Ardea 86.1 (1998): 1-10.

Manx Shearwater key prey species: herring (*Clupea harengus*), sprat (*Sprattus sprattus*), sardines (*Sardina pilchardus*), anchovies (*Engraulis encrasicolus*) and sandeels (*Ammodytes marinus*). This species also heavily preys on cephalopod species, however these are poorly characterised so were not included in this analysis. Prey species list was compiled from:

Riou, S., et al. "Recent impacts of anthropogenic climate change on a higher marine predator in western Britain." Marine Ecology Progress Series 422 (2011): 105-112.

Lee, D. S., et al. "Manx Shearwater (Puffinus puffinus), version 1.0." In Birds of the World (S. M. Billerman, Editor). Cornell Lab of Ornithology, Ithaca, NY, USA (2020).

Prey range information for all species were compiled from:

Kesner-Reyes, K., et al. "AquaMaps: Predicted range maps for aquatic species." In FishBase: R. Froese & D. Pauly (Eds.) (2019). Available at: https://www.aquamaps.org

Sailley, S., et al. "Fish abundance and catch data for the Northwest European Shelf and Mediterranean Sea from 2006 to 2098 derived from climate projections". Copernicus Climate Change Service (C3S) Climate Data Store (CDS) (2021). https://doi.org/10.24381/cds.39c97304.

1.1.4 Climate change impacts outside of Europe
Leach's Storm-petrel

Leach's storm-petrels in North America have changed their prey species and foraging strategy in response to shifts in the marine ecosystem partially driven by climate change. Heatwaves in North America have impacted storm-petrel colonies and resulted in changes in diet, loss of condition and wrecks. While individual heatwaves are difficult to attribute to climate change, it is likely the frequency and intensity of such events is increasing. Leach's storm-petrel reproductive success in Canada has been linked to global temperature. Warmer temperatures result in higher reproductive success, up until a certain threshold after which it decreases.

The mechanism is unknown.

Hedd, A., et al. "Diets and distributions of Leach's storm-petrel (Oceanodroma leucorhoa) before and after an ecosystem shift in the Northwest Atlantic." Canadian Journal of Zoology 87.9 (2009): 787-801.

D'Entremont, K., et al. "On-land foraging by Leach's Storm Petrels Oceanodroma leucorhoa coincides with anomalous weather conditions." Marine Ornithology 49 (2021): 247-252.

Mauck, R. A., Dearborn, D. C., and Huntington, C. E. "Annual global mean temperature explains reproductive success in a marine vertebrate from 1955 to 2010." Global Change Biology 24.4 (2018): 1599-1613.

Manx Shearwater

Manx shearwaters are known to be sensitive to climate change in the tropics, particularly to wrecks caused by storms, which are becoming more common due to changes in the El Nino cycle

Tavares, D. C., et al. "Mortality of seabirds migrating across the tropical Atlantic in relation to oceanographic processes." Animal Conservation 23.3 (2020): 307-319.

1.2 Sensitivity (references)

We used a list of candidate traits based on that in Foden & Young (2016) that indicate high sensitivity and identified which, if any, petrels and shearwaters possessed. In brief, we consulted published literature as well as expert knowledge and online databases such as Birdlife (http://datazone.birdlife.org/) and Birds of the World (https://birdsoftheworld.org), to assess whether petrels and shearwaters have either 1) Specialised habitat and/or microhabitat requirement 2) Environmental tolerances or thresholds (at any life stage) that are likely to be exceeded due to climate change 3) Dependence on environmental triggers that are likely to be disrupted by climate change, 4) Dependence on interspecific interactions that are likely to be disrupted by climate change or 5) High rarity. For more detail and a full list of traits see:

Foden, W. B. and Young, B. E. (eds.). "IUCN SSC Guidelines for Assessing Species' Vulnerability to Climate Change. Version 1.0." Occasional Paper of the IUCN Species Survival Commission No. 59 (2016). Cambridge, UK and Gland, Switzerland: IUCN Species Survival Commission. x+114pp.

1.3 Adaptive capacity (references)

We used a list of candidate traits based on that in Foden & Young (2016) that indicate adaptive capacity and identified which, if any, petrels and shearwaters possessed. In brief, we consulted published literature as well as expert knowledge and online databases such as Birdlife (http://datazone.birdlife.org/) and Birds of the World (https://birdsoftheworld.org), to assess whether petrels and shearwaters have either: 1) High phenotypic plasticity. 2) High dispersal ability or 3) High evolvability.

For more detail and a full list of traits see:

Foden, W. B. and Young, B. E. (eds.). "IUCN SSC Guidelines for Assessing Species' Vulnerability to Climate Change. Version 1.0." Occasional Paper of the IUCN Species Survival Commission No. 59 (2016). Cambridge, UK and Gland, Switzerland: IUCN Species Survival Commission. x+114pp.

© Seppo Häkkinen

Appendix 1: Skuas
Sources and references for vulnerability assessment

1.1 Evidence for exposure (references)

1.1.1 Current impacts attributed to climate change:

Great Skua

1 - Hotter summers result in increased heat stress in adults and chicks. Adults more frequently leave nests unattended to thermoregulate, which exacerbates chick heat stress

> Oswald, S. A., et al. "Heat stress in a high-latitude seabird: effects of temperature and food supply on bathing and nest attendance of great skuas Catharacta skua." Journal of Avian Biology 39.2 (2008): 163-169. In hot summers prey availability is typically lower, so nesting adults need to leave the nest to forage for longer periods of time. They also bath more regularly to thermoregulate. This results in lower fledgling rates as chicks are unattended for longer and left vulnerable to heat stress. Study was conducted on Foula, Shetlands. Note that study does not investigate a trend in temperature over its 3 study years, but presents evidence that temperature increases and prey shortages are linked and are driven by climate change

2 - In hotter summers, adults more frequently leave nests unattended due to prey shortages and to thermoregulate, which results in higher chick mortality due to predation

> Oswald, S. A., et al. "Heat stress in a high-latitude seabird: effects of temperature and food supply on bathing and nest attendance of great skuas Catharacta skua." Journal of Avian Biology 39.2 (2008): 163-169. In hot summers prey availability is typically lower, so nesting adults need to leave the nest to forage for longer periods of time. They also bathe more regularly to thermoregulate. This results in lower fledgling rates as chicks are unattended for longer and left vulnerable to predation. Study was conducted on Foula, Shetlands. Note that study does not investigate a trend in temperature over its 3 study years, but presents evidence that temperature increases and prey shortages are linked and are driven by climate change.

3 - Changes in prey availability during the breeding season have led to decreased fledgling success

> Oswald, S. A., et al. "Heat stress in a high-latitude seabird: effects of

temperature and food supply on bathing and nest attendance of great skuas Catharacta skua." Journal of Avian Biology 39.2 (2008): 163-169. Lower food availability (particularly sandeels) means adults must forage for longer, resulting in the same problem as for heat stress: chicks are left unattended and vulnerable. Study on Foula, Shetlands. Note that study does not investigate a trend in temperature over its 3 study years, but presents evidence that temperature increases and prey shortages are linked and are driven by climate change.

4 - Changes in prey availability has led to increased population size

Descamps, S., et al. "Climate change impacts on wildlife in a High Arctic archipelago–Svalbard, Norway." Global Change Biology 23.2 (2017): 490-502. Great skuas were first observed breeding on Svalbard in 1970, and their numbers have increased rapidly in recent years. This is most likely driven by range shifts in prey species and because of general warming of the climate.

Long-tailed Jaeger

1 - Southern populations are becoming less populous or going extinct in correlation with rising temperatures. Exact mechanism unknown, probably related to prey availability or heat stress.

Virkkala, R. and Rajasärkkä, A. "Northward density shift of bird species in boreal protected areas due to climate change." Boreal Environment Research 16 (suppl. B) (2011): 2–13. Long-tailed skuas have drastically reduced in range and density in Finland due to climate change, density in 2000-2009 was <50% of what it was in censuses carried out 1981–1999. Seems to be part of a range shift north. Exact mechanism unknown, probably related to prey availability or heat stress.

Arctic Jaeger

1 - Changes in prey availability have led to declines in key seabird species that Arctic skuas parasitise, thus leading to population declines in skuas.

Perkins, A., et al. "Combined bottom-up and top-down pressures drive catastrophic population declines of Arctic skuas in Scotland." Journal of Animal Ecology 87.6 (2018): 1573-1586. Arctic skuas in Scotland are declining drastically, there are multiple potential causes behind this. One likely driver is the decline of other seabirds due to climate change, which are important sources of food for skuas (usually by stealing their prey). Study looks at multiple colonies across Shetlands and Orkney Islands.

2 - Increased competition and predation from Great skuas, due to an increasing population size and prey swapping.

Perkins, A., et al. "Combined bottom-up and top-down pressures drive catastrophic population declines of Arctic skuas in Scotland." Journal of Animal Ecology 87.6 (2018): 1573-1586. Great skuas are increasing in number and have swapped their diet from predominantly fish to predominantly preying on other birds. Greater numbers of great skua lead to lower fledgling survival rate in arctic skuas. Study looks at multiple colonies across Shetlands and Orkney Islands.

Dawson, N. M., et al. "Interactions with Great Skuas Stercorarius skua as a factor in the long-term decline of an Arctic Skua Stercorarius parasiticus population." Ibis 153.1 (2011): 143-153. Arctic skua range is contracting across the Shetlands as great skua populations grow and expand. The overall decline in Arctic skua seems to be driven by this and lower density of sandeels, and both patterns are partially driven by climate change.

1.1.2 Change in European range size between present day and 2100:

Using a species distribution model (SDM) we correlated species occurrence during the breeding season with a number of terrestrial and marine environmental variables. Species range data came from the European Breeding Bird Atlas (EBBA2) database. Present-day and 2100 terrestrial data were downloaded from the WorldClim database. We used data from the MRI-ESM2 general circulation model (GCM), which is a high-performing model over Europe. Present-day and 2100 marine data were downloaded from the Bio-Oracle database which averages predictions of marine variables from several different atmospheric-oceanic general circulation models (AOGCMS; for full details see Assis et al., 2017). For the map presented in the summary we used representative concentration pathway (RCP) 4.5, which is an "intermediate" emissions scenario. All data were at 5-minute resolution.

For great skua and Arctic jaeger we included the following terrestrial variables: Mean temperature of the warmest month, precipitation during breeding season, isolation of landmass, area of landmass, distance to sea.

For long-tailed jaeger we included the following terrestrial variables: mean temperature of the warmest month, precipitation during breeding season, distance to sea.

For great skua and Arctic jaeger we included the following marine variables: sea surface temperature (during the winter), salinity, maximum chlorophyll concentration, bathymetry (depth and variance).

After running our model we generated a present-day map where every grid-cell is given a habitat suitability score between 0 and 1, where 1 is very suitable habitat and 0 is not at all suitable. We then compared this with a corresponding map built with 2100 data, and highlighted currently inhabited areas where 1) suitability drops sharply (i.e. by more than 0.1) and 2) suitability drops below a probability threshold set by the model. Conversely we also highlighted areas where suitability rose sharply and above a given threshold. While a drop in habitat suitability is likely to result in population declines, it is not a certainty, and it does not mean that a population will be extinct in 2100 or that a population is doomed to extinction. With conservation action and careful management, along with changes in human behaviour, such declines may be mitigated or in some cases prevented. For a full explanation of the model see the accompanying 'Methodology' folder in Appendix 2.

Underlying data were downloaded from:

Keller, V., et al. "European Breeding Bird Atlas 2: Distribution, Abundance and Change." European Bird Census Council & Lynx Edicions, Barcelona (2020). Source of range data

Fick, S. E., and Hijmans, R. J. "WorldClim 2: new 1-km spatial resolution climate surfaces for global land areas." International Journal of Climatology 37.12 (2017): 4302-4315. Source of present-day and 2100 terrestrial data.

Assis, J., et al. "Bio-ORACLE v2. 0: Extending marine data layers for bioclimatic modelling." Global Ecology and Biogeography 27.3 (2018): 277-284. Source of present-day and 2100 marine data

1.1.3 Changes in key prey species:

We first identified the key prey species for each species. This can be variable across a species' range, but if available evidence suggested at least one major population is highly dependent on a particular prey species, then typically this species would be included. Lists of prey species were compiled from published sources, then verified and expanded following consultation with conservation practitioners. Afterwards we compiled current and projected maps of prey ranges to assess where key prey species may become less common in the near future. If any of the key species are predicted to vanish or drastically reduce in abundance in the current foraging range a given species, we marked this on the summary map.

We used several sources to collate range information, but for preference we used data from COPERNICUS as they include projected abundance. For species where this was not available we used habitat suitability instead. In all cases we used RCP 4.5, which is an "intermediate" emissions scenario. For species in the COPERNICUS database we used the 0.6 maximum sustainable yield parameter, which assumes international co-operation to work towards fish-stock sustainability. Our assessment is therefore relatively conservative in terms of changes in prey species.

Great Skua key prey species: whiting (*Merlangius merlangus*), blue whiting (*Micromesistius poutassou*), haddock (*Melanogrammus aeglefinus*) and pout (*Trisopterus esmarkii*). Some populations rely heavily on predation of other seabirds (such as fulmars, kittiwakes and puffins) and on fishery discards. These were not included in our prey assessment. Prey species list was compiled from:

> Jones, T., et al. "Breeding performance and diet of Great Skuas Stercorarius skua and Parasitic Jaegers (Arctic Skuas) S. parasiticus on the west coast of Scotland." Bird Study 55.3 (2008): 257-266.

> Votier, S. C., et al. "Predation by great skuas at a large Shetland seabird colony." Journal of Applied Ecology 41.6 (2004): 1117-1128.

Long-tailed Jaeger key prey species: collared lemming (*Dichrostonyx groenlandicus*), gray-sided vole (*Clethrionomys rufocanus*) and Norway lemming (*Lemmus lemmus*). Breeding populations rely heavily on rodents, including collared lemmings, gray-sided voles and Norway lemmings. Where rodents are not common, some populations rely on kleptoparasitism. These species were not included in our prey assessment. Prey species list was compiled from:

> Dekorte, J., And Wattel, J. "Food and breeding success of the long-tailed skua at Scoresby Sund, Northeast Greenland." Ardea 76.1 (1988): 27-41.

> Andersson, M. "Population ecology of the long-tailed skua (Stercorarius longicaudus Vieill.)." The Journal of Animal Ecology (1976): 537-559.

Arctic Jaeger key prey species: sandeel species (*Ammodytes marinus* and *Ammodytes tobianus*). Prey species list was compiled from:

> Phillips, R. A., Caldow, R. W. G., and Furness, R. W. "The influence of food availability on the breeding effort and reproductive success of Arctic Skuas Stercorarius parasiticus." Ibis 138.3 (1996): 410-419.

Prey range information for all species were compiled from:

> Kesner-Reyes, K., et al. "AquaMaps: Predicted range maps for aquatic species." In FishBase: R. Froese & D. Pauly (Eds.) (2019). Available at: https://www.aquamaps.org

> Sailley, S., et al. "Fish abundance and catch data for the Northwest European Shelf and Mediterranean Sea from 2006 to 2098 derived from climate projections". Copernicus Climate Change Service (C3S) Climate

Data Store (CDS) (2021). https://doi.org/10.24381/cds.39c97304.

1.1.4 Climate change impacts outside of Europe
Long-tailed Jaeger

Skuas have been heavily affected by climate change in Greenland, in particular due to lack of prey and increased predation due to other species prey-switching

Schmidt, N. M., et al. "Response of an arctic predator guild to collapsing lemming cycles." Proceedings of the Royal Society B: Biological Sciences 279.1746 (2012): 4417-4422.

Barraquand, F., et al. "Demographic responses of a site-faithful and territorial predator to its fluctuating prey: long-tailed skuas and arctic lemmings." Journal of Animal Ecology 83.2 (2014): 375-387.

1.2 Sensitivity (references)

We used a list of candidate traits based on that in Foden & Young (2016) that indicate high sensitivity and identified which, if any, skuas possessed. In brief, we consulted published literature as well as expert knowledge and online databases such as Birdlife (http://datazone.birdlife.org/) and Birds of the World (https://birdsoftheworld.org), to assess whether skuas have either 1) Specialised habitat and/or microhabitat requirement 2) Environmental tolerances or thresholds (at any life stage) that are likely to be exceeded due to climate change 3) Dependence on environmental triggers that are likely to be disrupted by climate change, 4) Dependence on interspecific interactions that are likely to be disrupted by climate change or 5) High rarity.

For more detail and a full list of traits see:

Foden, W. B. and Young, B. E. (eds.). "IUCN SSC Guidelines for Assessing Species' Vulnerability to Climate Change. Version 1.0." Occasional Paper of the IUCN Species Survival Commission No. 59 (2016). Cambridge, UK and Gland, Switzerland: IUCN Species Survival Commission. x+114pp.

1.3 Adaptive capacity (references)

We used a list of candidate traits based on that in Foden & Young (2016) that indicate adaptive capacity and identified which, if any, skuas possessed. In brief, we consulted published literature as well as expert knowledge and online databases such as Birdlife (http://datazone.birdlife.org/) and Birds of the World (https://birdsoftheworld.org), to assess whether skuas have either: 1) High phenotypic plasticity. 2) High dispersal ability or 3) High evolvability.

For more detail and a full list of traits see:

Foden, W. B. and Young, B. E. (eds.). "IUCN SSC Guidelines for Assessing Species' Vulnerability to Climate Change. Version 1.0." Occasional Paper of the IUCN Species Survival Commission No. 59 (2016). Cambridge, UK and Gland, Switzerland: IUCN Species Survival Commission. x+114pp.

© Seppo Häkkinen

Appendix 1: Terns
Sources and references for vulnerability assessment

1.1 Evidence for exposure (references)

1.1.1 Current impacts attributed to climate change:

Arctic Tern

1 - Changes in prey availability during the breeding season have led to population declines (debated).

> Lindegren, M., et al. "Productivity and recovery of forage fish under climate change and fishing: North Sea sandeel as a case study." Fisheries Oceanography 27.3 (2018): 212-221. Severe decline in sandeels, which Arctic terns are heavily dependent on in the UK, has led to population declines across the Shetlands. The decline in sandeels is likely linked to fisheries rather than climate change, but it is believed climate change may be hindering recovery of sandeel populations. Sandeel abundance is lower in warmer temperatures, so it is plausible that climate change will have a negative effect on sandeel availability.

2 - Changes in prey availability during the breeding season have led to population declines.

> Vigfusdottir, F. "Drivers of productivity in a subarctic seabird: Arctic Terns in Iceland." University of East Anglia (2012). PhD Dissertation. Increasing sea temperatures around Iceland have been linked to reduced sandeel recruitment, which is the likely cause behind recent population declines
> Petersen, A., et al. "Annual survival of Arctic terns in western Iceland." Polar Biology 43.11 (2020): 1843-1849. Updated evidence supporting the above study, Arctic terns in Iceland are declining and likely cause is collapse of sandeel populations related to climate change

3 - Arctic terns are arriving from migration and breeding earlier

> Wanless, S., et al. "Long-term changes in breeding phenology at two seabird colonies in the western North Sea." Ibis 151.2 (2009): 274-285. Arctic terns on the Farne Islands are arriving and breeding earlier. Study used a 35-year dataset. The study suggests climatic conditions in wintering grounds or during the spring migration may have driven phenological change.
> Møller, A. P., Flensted-Jensen, E., and Mardal, W. "Rapidly advancing laying date in a seabird and the changing advantage of early reproduction." Journal of Animal Ecology (2006): 657-665. Arctic terns in Demark are

arriving and breeding earlier. Strong correlation with changes in NAO and temperature.

4 - Juvenile Arctic terns have begun to disperse further, distance has increased in correlation with warmer winters. Mechanism unknown, but likely mediated through prey availability

> **Møller, A. P., Flensted-Jensen, E., and Mardal, W. "Dispersal and climate change: a case study of the Arctic tern Sterna paradisaea." Global Change Biology 12.10 (2006): 2005-2013.** Between 1933 and 1997, the mean natal dispersal distance of Danish Arctic terns increased from around 10 km to around 100 km. (Natal dispersal is the movement from birth site to first breeding site.) The winter NAO index increased over the same period. The study suggests the wNAO may affect dispersal decision through effects on the marine ecosystem, and that increased natal dispersal distances have a fitness cost because they delay breeding and therefore reduce recruitment probability.

Little Tern

1 - Little tern nests are frequently washed away by tidal surges, such events are becoming more frequent or extensive due to rising sea levels

> **Johnson, C., Sullivan, I., and Newton, S. "Tern Colony Management and Protection at Kilcoole 2017." Department of Arts, Heritage and the Gaeltacht: BirdWatch Ireland (2017).** Little tern nests in the UK are frequently washed away by tidal surges (personal correspondence), which has strongly contributed to successive years of poor breeding and subsequent population decline in the UK. Rising sea level exacerbates these effects and reduces the amount of safe breeding habitat available.

2 - As sea temperature has increased over time, tern productivity has decreased. Mechanism unknown, but likely mediated through prey availability

> **Burthe, S. J., et al. "Assessing the vulnerability of the marine bird community in the western North Sea to climate change and other anthropogenic impacts." Marine Ecology Progress Series 507 (2014): 277-295.** Tern productivity is lower when the sea surface temperature is higher. Mechanism probably linked to prey availability. Local sea surface temperature increased over the study duration. Study focusses on seabirds in Forth and Tay region.

Sandwich Tern

1 - Sandwich terns are changing their migration timing and arriving earlier to breeding sites

> **Wanless, S., et al. "Long-term changes in breeding phenology at two**

seabird colonies in the western North Sea." Ibis 151.2 (2009): 274-285.
Sandwich terns in the Farne Islands are arriving earlier, likely in response to
changing environmental cues in wintering grounds or on the spring migration
route. However, there was no significant advancement in laying date. Study
used a roughly 30-year dataset.

2 - Sandwich terns are changing their migration timing, both migration and
breeding events are occurring later, making the breeding season shorter

**Møller, A. P., et al. "Climate change affects the duration of the reproductive
season in birds." Journal of animal ecology 79.4 (2010): 777-784.** Breeding
season is now significantly shorter in Denmark than it was in 1970 (by approx.
36 days). This is in correlation with rising spring temperatures, is likely in
response to changes in environmental conditions

1.1.2 Change in European range size between present day and 2100:

Using a species distribution model (SDM) we correlated species occurrence during
the breeding season with a number of terrestrial and marine environmental
variables. Species range data came from the European Breeding Bird Atlas (EBBA2)
database. Present-day and 2100 terrestrial data were downloaded from the
WorldClim database. We used data from the MRI-ESM2 general circulation model
(GCM), which is a high-performing model over Europe. Present-day and 2100
marine data were downloaded from the Bio-Oracle database which averages
predictions of marine variables from several different atmospheric-oceanic general
circulation models (AOGCMS; for full details see Assis et al., 2017). For the map
presented in the summary we used representative concentration pathway (RCP)
4.5, which is an "intermediate" emissions scenario. All data were at 5-minute
resolution.

For Caspian tern, roseate tern, Arctic tern, little tern, and Sandwich tern we
included the following terrestrial variables: Mean temperature of the warmest
month, precipitation during breeding season, isolation of landmass, area of
landmass, distance to sea.

For Caspian tern, roseate tern, little tern, and Sandwich tern we included the
following marine variables: sea surface temperature (during the winter), salinity,
maximum chlorophyll concentration, bathymetry (depth and variance)

After running our model we generated a present-day map where every grid-cell is
given a habitat suitability score between 0 and 1, where 1 is very suitable habitat
and 0 is not at all suitable. We then compared this with a corresponding map built
with 2100 data, and highlighted currently inhabited areas where 1) suitability drops

sharply (i.e. by more than 0.1) and 2) suitability drops below a probability threshold set by the model. Conversely we also highlighted areas where suitability rose sharply and above a given threshold. While a drop in habitat suitability is likely to result in population declines, it is not a certainty, and it does not mean that a population will be extinct in 2100 or that a population is doomed to extinction. With conservation action and careful management, along with changes in human behaviour, such declines may be mitigated or in some cases prevented. For a full explanation of the model see the accompanying 'Methodology' folder in Appendix 2.

Underlying data were downloaded from:

Keller, V., et al. "European Breeding Bird Atlas 2: Distribution, Abundance and Change." European Bird Census Council & Lynx Edicions, Barcelona (2020). Source of range data

Fick, S. E., and Hijmans, R. J. "WorldClim 2: new 1-km spatial resolution climate surfaces for global land areas." International Journal of Climatology 37.12 (2017): 4302-4315. Source of present-day and 2100 terrestrial data.

Assis, J., et al. "Bio-ORACLE v2. 0: Extending marine data layers for bioclimatic modelling." Global Ecology and Biogeography 27.3 (2018): 277-284. Source of present-day and 2100 marine data

1.1.3 Changes in key prey species:
We first identified the key prey species for each species. This can be variable across a species' range, but if available evidence suggested at least one major population is highly dependent on a particular prey species, then typically this species would be included. Lists of prey species were compiled from published sources, then verified and expanded following consultation with conservation practitioners. Afterwards we compiled current and projected maps of prey ranges to assess where key prey species may become less common in the near future. If any of the key species are predicted to vanish or drastically reduce in abundance in the current foraging range a given species, we marked this on the summary map.
We used several sources to collate range information, but for preference we used data from COPERNICUS as they include projected abundance. For species where this was not available we used habitat suitability instead. In all cases we used RCP 4.5, which is an "intermediate" emissions scenario. For species in the COPERNICUS database we used the 0.6 maximum sustainable yield parameter, which assumes international co-operation to work towards fish-stock sustainability. Our assessment is therefore relatively conservative in terms of changes in prey species.

Arctic Tern key prey species: sandeel species (*Ammodytes marinus* and

Ammodytes tobianus), herring (*Clupea harengus*) and stickleback (*Gasterosteus aculeatus*). This species also preys on small invertebrates (larval fish, shrimp, idotea, etc.). These are poorly characterised so have not been included in the prey assessment. Prey species list was compiled from:

Hatch, J. J., et al. "Arctic Tern (Sterna paradisaea), version 1.0." In Birds of the World (S. M. Billerman, Editor). Cornell Lab of Ornithology, Ithaca, NY, USA (2020).

Eglington, S., and Perrow, M. R. "Literature review of tern (Sterna & Sternula spp.) foraging ecology". ECON Ecological Consultancy Limited (2014).

Little Tern key prey species: sandeel species (*Ammodytes marinus* and *Ammodytes tobianus*), herring (*Clupea harengus*), sprat (*Sprattus sprattus*), sand-smelt (*Atherina presbyter*), sardines (*Sardina pilchardus*) and goby species. Prey species list was compiled from:

Shealer, D., et al. "Sandwich Tern (Thalasseus sandvicensis), version 1.0." In Birds of the World (S. M. Billerman, Editor). Cornell Lab of Ornithology, Ithaca, NY, USA (2020).

Eglington, S., and Perrow, M. R. "Literature review of tern (Sterna & Sternula spp.) foraging ecology". ECON Ecological Consultancy Limited (2014).

Roseate Tern key prey species: sprat (*Sprattus sprattus*) and sandeel species (*Ammodytes marinus* and *Ammodytes tobianus*). Prey species list was compiled from:

Green, E. "Tern diet in the UK and Ireland: a review of key prey species and potential impacts of climate change." RSPB Report (2017).

Eglington, S., and Perrow, M. R. "Literature review of tern (Sterna & Sternula spp.) foraging ecology". ECON Ecological Consultancy Limited (2014).

Sandwich Tern key prey species: sandeel species (*Ammodytes marinus* and *Ammodytes tobianus*), herring (*Clupea harengus*), sprat (*Sprattus sprattus*), anchovy (*Engraulis encrasicholus*) and sardine (*Sardina pilchardus*). Prey species list was compiled from:

Shealer, D., et al. "Sandwich Tern (Thalasseus sandvicensis), version 1.0." In Birds of the World (S. M. Billerman, Editor). Cornell Lab of Ornithology, Ithaca, NY, USA (2020).

Eglington, S., and Perrow, M. R. "Literature review of tern (Sterna & Sternula spp.) foraging ecology". ECON Ecological Consultancy Limited (2014).

Caspian Tern key prey species: herring (*Clupea harengus*). Prey species list was compiled from:

Cuthbert, F. J. and Wires, L. R. "Caspian Tern (Hydroprogne caspia), version 1.0." In Birds of the World (S. M. Billerman, Editor). Cornell Lab of Ornithology, Ithaca, NY, USA (2020).

Prey range information for all species were compiled from:

Kesner-Reyes, K., et al. "AquaMaps: Predicted range maps for aquatic species." In FishBase: R. Froese & D. Pauly (Eds.) (2019). Available at: https://www.aquamaps.org

Sailley, S., et al. "Fish abundance and catch data for the Northwest European Shelf and Mediterranean Sea from 2006 to 2098 derived from climate projections". Copernicus Climate Change Service (C3S) Climate Data Store (CDS) (2021). https://doi.org/10.24381/cds.39c97304.

1.1.4 Climate change impacts outside of Europe

Caspian Tern

Caspian terns in North America have been negatively affected by heatwaves, warming seas, severe storms, and increased frequency of flooding, all of which are linked to climate change.

Suzuki, Y., et al. "Colony connectivity and the rapid growth of a Caspian Tern (Hydroprogne caspia) colony on Alaska's Copper River Delta, USA." Waterbirds 42.1 (2019): 1-7.

1.2 Sensitivity (references)

We used a list of candidate traits based on that in Foden & Young (2016) that indicate high sensitivity and identified which, if any, terns possessed. In brief, we consulted published literature as well as expert knowledge and online databases such as Birdlife (http://datazone.birdlife.org/) and Birds of the World (https://birdsoftheworld.org), to assess whether terns have either 1) Specialised habitat and/or microhabitat requirement 2) Environmental tolerances or thresholds (at any life stage) that are likely to be exceeded due to climate change 3) Dependence on environmental triggers that are likely to be disrupted by climate change, 4) Dependence on interspecific interactions that are likely to be disrupted by climate change or 5) High rarity.

For more detail and a full list of traits see:

> Foden, W. B. and Young, B. E. (eds.). "IUCN SSC Guidelines for Assessing Species' Vulnerability to Climate Change. Version 1.0." Occasional Paper of the IUCN Species Survival Commission No. 59 (2016). Cambridge, UK and Gland, Switzerland: IUCN Species Survival Commission. x+114pp.

1.3 Adaptive capacity (references)

We used a list of candidate traits based on that in Foden & Young (2016) that indicate adaptive capacity and identified which, if any, terns possessed. In brief, we consulted published literature as well as expert knowledge and online databases such as Birdlife (http://datazone.birdlife.org/) and Birds of the World (https://birdsoftheworld.org), to assess whether terns have either: 1) High phenotypic plasticity. 2) High dispersal ability or 3) High evolvability.

For more detail and a full list of traits see:

> Foden, W. B. and Young, B. E. (eds.). "IUCN SSC Guidelines for Assessing Species' Vulnerability to Climate Change. Version 1.0." Occasional Paper of the IUCN Species Survival Commission No. 59 (2016). Cambridge, UK and Gland, Switzerland: IUCN Species Survival Commission. x+114pp.

© Seppo Häkkinen

275 Climate Change Vulnerability and Conservation: Seabirds

Appendix 2: Sources and references for conservation action assessment

Due to the volume of data involved, we have placed our supporting information detailing our full methodology and list of references into a separate resource. It is available to download here: https://doi.org/10.11647/OBP.0343.

Methodology

If you would like to see our full methodology, please see the accompanying "Methodology" folder. This folder contains information on:

- How we compiled our climate change impacts list for all species

- How we compiled our conservation action evidence base

- How we assessed effectiveness of conservation actions

- How we scored strength, transparency and relevance of conservation actions

- Additional methodology on modelling approach and generating outputs

References

If you would like a full reference list of all studies in our evidence base, as well as the raw data for the evidence base please see the "References" folder. This folder contains:

- All references relevant to climate change impacts, and each species' sensitivity and adaptive capacity to climate change

- All references used to assess species diet

- The full citation of each paper included in the conservation action evidence base and additional details. Additional details include the focal species of each study, the sample size and location of each study, what the metric of success was for each study, whether each study had published and replicable methodology, whether we considered that each study had a clear justification of their method, whether we considered that each study had a clear outcome, the form of each study, and each study's experimental design

- Links to relevant Conservation Evidence website data (where applicable)

- Additional details on ex-situ populations, rehabilitation and release.

Version History

This guidance is continually updated, we include a brief version history here to inform readers on changes between versions. This version is v1.2. If you would like to see the full online version please visit this address: https://doi.org/10.11647/ OBP.0343.

Version 1:

v1.0: Full initial release.

v1.1: Minor fixes. Error found and now fixed on ivory gull vulnerability map. Additional labels added to photos. Version history page added.

v1.2: Information from EAZA Charadriiformes TAG incorporated. Includes additional information on seabird ex-situ population sizes, breeding success, rehabilitation and release, and overview of head-starting for different species.

© Seppo Häkkinen